The Open University

GW00640935

MU120
Open Mathematics

Unit 14

Space and shape

MU120 course units were produced by the following team:

Gaynor Arrowsmith (Course Manager)
Mike Crampin (Author)
Margaret Crowe (Course Manager)
Fergus Daly (Academic Editor)
Judith Daniels (Reader)
Chris Dillon (Author)
Judy Ekins (Chair and Author)
John Fauvel (Academic Editor)
Barrie Galpin (Author and Academic Editor)
Alan Graham (Author and Academic Editor)
Linda Hodgkinson (Author)
Gillian Iossif (Author)
Joyce Johnson (Reader)
Eric Love (Academic Editor)
Kevin McConway (Author)
David Pimm (Author and Academic Editor)
Karen Rex (Author)

Other contributions to the text were made by a number of Open University staff and students and others acting as consultants, developmental testers, critical readers and writers of draft material. The course team are extremely grateful for their time and effort.

The course units were put into production by the following:

Course Materials Production Unit (Faculty of Mathematics and Computing)

Martin Brazier (Graphic Designer)	Diane Mole (Graphic Designer)
Hannah Brunt (Graphic Designer)	Kate Richenburg (Publishing Editor)
Alison Cadle (TEXOpS Manager)	John A.Taylor (Graphic Artist)
Jenny Chalmers (Publishing Editor)	Howie Twiner (Graphic Artist)
Sue Dobson (Graphic Artist)	Nazlin Vohra (Graphic Designer)
Roger Lowry (Publishing Editor)	Steve Rycroft (Publishing Editor)

This publication forms part of an Open University course. Details of this and other Open University courses can be obtained from the Student Registration and Enquiry Service, The Open University, PO Box 197, Milton Keynes MK7 6BJ, United Kingdom: tel. +44 (0)845 300 6090, email general-enquiries@open.ac.uk

Alternatively, you may visit the Open University website at http://www.open.ac.uk where you can learn more about the wide range of courses and packs offered at all levels by The Open University.

To purchase a selection of Open University course materials visit http://www.ouw.co.uk, or contact Open University Worldwide, Walton Hall, Milton Keynes MK7 6AA, United Kingdom, for a brochure: tel. +44 (0)1908 858793, fax +44 (0)1908 858787, email ouw-customer-services@open.ac.uk

The Open University, Walton Hall, Milton Keynes, MK7 6AA.

First published 1996. Second edition 2002. Third edition 2008.

Copyright © 1996, 2002, 2008 The Open University

Edited, designed and typeset by The Open University, using the Open University TEX System.

Printed and bound in the United Kingdom by The Charlesworth Group, Wakefield.

ISBN 978 0 7492 2871 2

3.1

Contents

Study guide

Unit 14 marks the beginning of the final block. It builds upon what you learned about maps in *Unit 6*, and also upon the trigonometric ideas you met in *Unit 9*. The unit is about space and shape, which are very visual topics, so there is considerable use of video material. There is also a short audio band and, as usual, some work from the *Calculator Book*.

This unit uses quite a number of geometric concepts that are covered in Module 7 of *Preparatory Resource Book B*. You may need to refer back to that module if you have forgotten some of the terms that crop up. The unit also makes use of ideas about proportion from *Unit 13*.

There are five main sections in the unit:

Section 1 revisits the concept of similarity, which was discussed in *Unit 2* and which underlies the idea of map scales in *Unit 6*. So you may like to refer to any notes that you made on similarity and scale when you studied *Units 2* and *6*. There is also a short extract from a Sherlock Holmes' story in the Reader.

Section 2 uses the concept of similarity to explain perspective drawings, which are essentially two-dimensional representations of three-dimensional shapes. There is an associated video band, entitled 'Getting things into perspective'. There are also two Reader articles to peruse, as well as Section 14.3 of the *Calculator Book* to study (note that this section is studied out of order).

Section 3 builds upon ideas of similarity in developing concepts that underpin the trigonometric functions you met in *Unit 9*.

Section 4 shows the application of trigonometry in map making and other types of surveying. There is a short audio band, and a related *Calculator Book* section.

Section 5 illustrates the use of trigonometry in navigation. It includes a video band, 'Showing the way they went'. Part 1 of the band looks back at the making of the *Unit 6* video band about planning a walk. You will need the OS map from *Unit 6* and any relevant notes that you made while studying it. Part 2 of the video looks at the use of trigonometry in the navigation of the globe. This section also involves studying Section 14.2 of the *Calculator Book*.

There is an *optional* Appendix that derives some of the trigonometric formulas discussed in the main text.

The television programme *Building by Numbers* is relevant to this unit.

As usual there are Handbook sheets associated with the unit, on which you can record any new terms or techniques as you encounter them.

If you are short of time, then Sections 1 and 3 are by far the most important ones; note that some activities are optional, as is the Appendix.

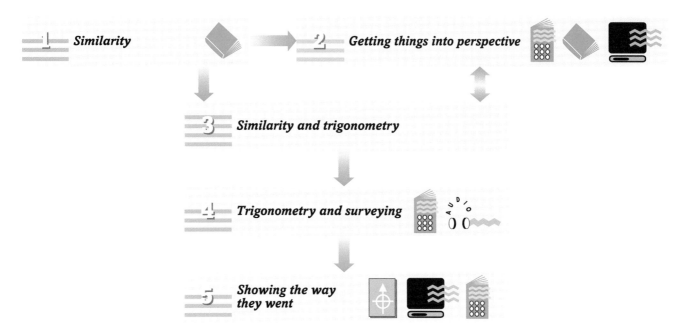

Summary of sections and other course components needed for *Unit 14*.

Introduction to Block D

This final block of the course is called *Sight and Sound.* It investigates mathematically the sight of pictures, maps and rainbows, and the sound of musical notes.

Unit 14 focuses on maps and pictures. Geometric objects (such as lines, triangles and circles) and trigonometric functions (which you met in *Unit 9*) are used to explain how maps, perspective drawings and other mathematical models of the world are produced.

Unit 15 revisits the discussion of musical sound from *Unit 9*, but offers a more detailed mathematical perspective on the characteristics of sounds of various sorts. Trigonometric relationships come to the fore in this unit, building upon ideas from *Unit 14*.

While there are some new mathematical themes in this block, the overall aim is to consolidate your learning—this is particularly true of *Unit 16*.

All the units in this block draw upon ideas from several earlier units, as indicated in the diagram below. You may find it helpful to refer to these units whilst studying the block.

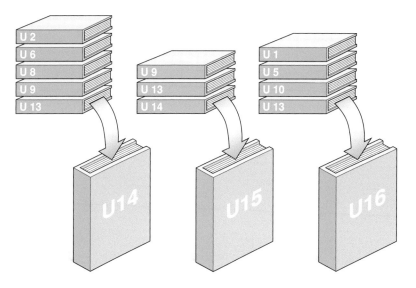

Links between units.

This block also gives you the opportunity to consider the ways in which you have been learning mathematics and what you have learned. One of the main aims of this course is that you should be able to use the knowledge you have acquired more widely. Identifying what you have done and how you have done it are ways of helping you to achieve this aim.

Introduction

The problems of representing a three-dimensional world on a two-dimensional map were discussed in *Unit 6*. A good map shows true *shape*, accurately scaled *distances* and *areas*, as well as consistent *orientation*. Paintings, like maps, provide two-dimensional representations of a three-dimensional world. However, in paintings, shapes are not generally preserved—a shape drawn 'in perspective' to 'look like' a square is often not itself a square!

When you attempt to reconstruct the three-dimensional world represented by a painting or a map, there is always some ambiguity. For example, you cannot be sure what size an object shown in a painting really is, or what the hillside really is like between two contour lines. Maps and pictures are drawn from particular points of view, with particular ways of seeing. As you work through this unit, bear in mind the notion of viewpoint and how it affects what is seen.

Maps are examples of *visual* models of the real world. This unit explores the use of geometric ideas in making visual models of real-world situations—from the art of drawing pictures in perspective to the technology of surveying for map-making.

One such visual model is the mathematical plane, which is a seamless expanse of uninterrupted flatness in all directions. Such planes provide the basis of many of the two-dimensional models of the real world that are discussed in this unit.

The mathematical plane.

As well as extending mathematical ideas, this unit (and the other units in the block) give you the opportunity to review the different strategies you have used so far in the course—for instance, how you take notes or tackle activities, or what you do when you are stuck.

Activity 1   *Your strategies*

For this activity, there is a cream-coloured activity sheet, entitled 'Strategies', that you may find useful.

The Study guide on page 4 outlines the concepts, skills and techniques from earlier in the course that you will be building upon in this unit. Consider your strategy for revising these ideas. Will you review the relevant unit sections or just your notes? Will you do that now or when the idea comes up in the text?

As you work through this unit, and the remainder of the block, note down the strategies you use and where each is useful. This should help in consolidating your approach to learning mathematics.

1 Similarity

Aims   This section aims to build upon some of the ideas concerning maps, similarity, angles, scale factors and ratios from earlier in the course.   ◇

You were introduced to the idea of similarity in *Unit 2*. For instance, the shapes in a photograph are similar to the corresponding shapes in an enlargement of that photograph: one is a scaled version of the other. Thus, figures are said to be *similar* when one figure is a scaled version of another: they are the same shape, although probably not the same size.

1.1 Similar figures

Figure 1 is part of the OS map used in *Unit 6*. It shows a number of fields, hills and other features.

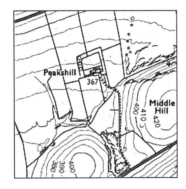

Figure 1   Part of OS map.

The map should provide a true representation of the shape of each of the actual fields (at the time the map was made and within the limits imposed by surveying, printing etc.). It does so because it is drawn to scale, and hence each field on the map is a scale representation of an actual field.

The scale of the map is 1 : 25 000. Recall from *Unit 6* that this means

   distance on the ground = 25 000 × corresponding distance on the map.

In general, the relationship between distances on a map and the corresponding distances on the ground is

   distance on the ground = scale factor × distance on the map.

From *Unit 13*, you should recognize this as a directly proportional relationship.

This means that when you measure a distance on the map, your measurement does not give the distance on the ground immediately. You need to carry out a calculation—specifically, you need to multiply the measurement by the scale factor, which in this case is 25 000.

▶ There is one kind of measurement that you can make on a map, which does give the corresponding measurement on the ground directly, without any processing. What is it?

The measurement of angles. Angles are identical on the map and on the ground—they are not scaled. When you measure the angle between two lines on a map with a compass scale or a protractor, your answer gives you the corresponding angle on the ground. For example, the angle on the map between the lines marking the road and the track is the same as the angle between the actual road and track.

Therefore the basic assumptions underlying the use of maps are:
- you can work out lengths on the ground by scaling the measurements of lengths on the map;
- you can find angles directly by measuring angles on the map.

These properties are specifically associated with shapes that are *similar*; in other words, shapes on maps are similar to the shapes on the ground that they represent.

Similarity is the mathematical formulation of the everyday notion of 'same shape'. As already mentioned, the idea of *similarity* involves shapes that are scaled versions of each other. Similar shapes can be created by enlargement. For instance, in Figure 2 the shapes, like the sage rectangles, parsley triangles and the basil circle, in the enlarged plan of the herb garden are similar to the corresponding shapes in the small plan.

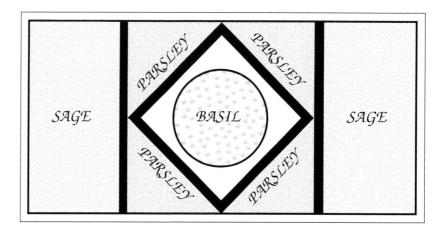

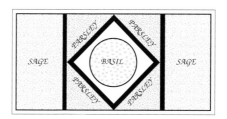

Figure 2   Similar plans of a herb garden.

The enlargement of any shape can be constructed easily. You need to decide a scale factor and choose a centre of enlargement. Draw lines from the centre to each of the vertices of the original shape and extend the lines, as in Figure 3. The distance of each vertex from the centre is then multiplied by the scale factor and marked out on the line through that vertex. This gives a new point on each line—these points are the vertices of the enlarged shape.

A vertex is a corner; vertices are corners.

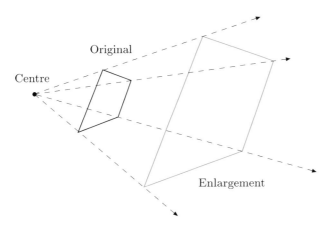

Figure 3   Constructing an enlargement.

A surprising feature is that, for a *given* scale factor, whatever centre you choose, the enlarged shape will always be the same size, but it will be in a different position according to the centre chosen. Each of the sides of the enlargement will always be parallel to the corresponding original side, and all the sides will be scaled by the same scale factor. This is true not only for the sides, but also for any other lines—for example, diagonals. However, the corresponding angles will all remain the same.

Where the scale factor is less than 1, the shape will shrink rather than enlarge. But, for convenience, this is also often described as an 'enlargement'.

Mathematical similarity is a slightly wider notion than enlargement. Similar shapes need not have corresponding sides parallel: they may be in different orientations, or even turned over, as in Figure 4.

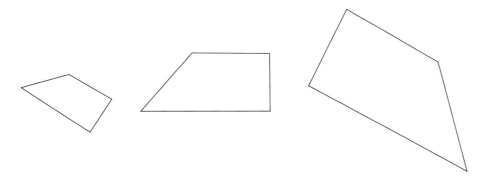

Figure 4   Similar shapes.

You may now find it useful to look back at any notes that you made about the word *similar* when you were studying *Unit 2*, and add to them.

Some kinds of shapes are always similar to each other: all squares are similar, all equilateral triangles are similar, and also all circles—as is apparent in Figure 5. The reasons for this are not difficult to see. With squares, all the angles are right angles, and since all the sides are the same length, they will still all be equal to one another when they are scaled up. In an equilateral triangle, all the angles are 60°, and all the sides are the same length and remain so when they are scaled up. For circles, only their size changes: there are no angles to consider because circles do not have vertices.

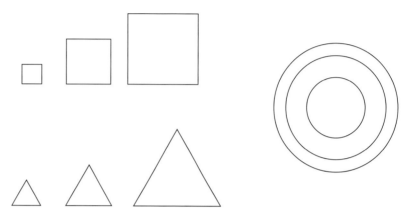

Figure 5   Families of similar shapes.

However, not all rectangles are similar: there are long thin rectangles as well as short fat ones. Figure 6(a) shows various non-similar rectangles, while Figure 6(b) shows a pair of similar rectangles.

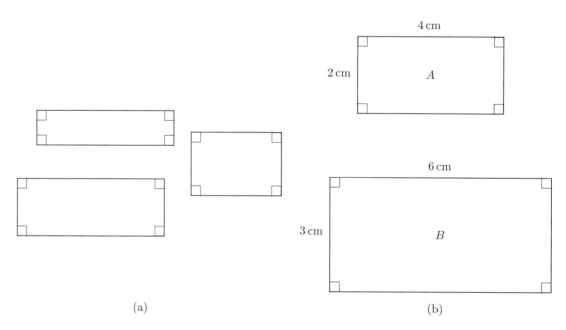

(a)   (b)

Figure 6   Non-similar and similar rectangles.

Look at Figure 6(b). The shapes are both rectangles (because each has four right angles). The rectangles are similar because one is a scaled-up version of the other: the lengths of the sides of the larger rectangle are each 1.5 times the corresponding sides of the smaller rectangle.

Activity 2   *Similar shapes*

Consider the pairs of shapes in Figure 7. Which pairs are similar?

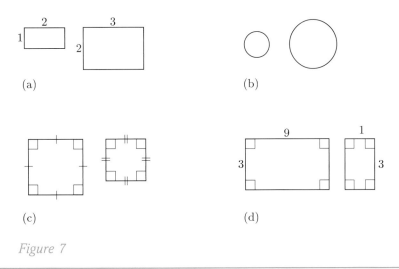

Figure 7

In addition to the scale factor of enlargement, there are other important relationships between pairs of similar figures. Look again at the two rectangles in Figure 6(b). Notice that not only are the rectangles similar with a scale factor of 1.5, but also that the ratio of the sides *within* one rectangle is the same as the ratio of the corresponding sides *within* the other rectangle: the length of the longer side is twice the length of the shorter side, as summarized in the table below.

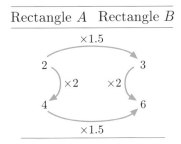

The lengths in pairs of similar shapes always have this dual aspect: they involve an enlargement *between* shapes, governed by the scale factor (1.5 in this case); and *within* each shape, corresponding lengths are in the same ratio. The scale factor is *between* the two shapes; the ratios are *within* each

13

shape. This gives two alternative ways of finding unknown lengths in a pair of similar shapes.

Example 1   *Finding unknown lengths*

The two triangles in Figure 8 are similar (triangle B is an enlargement of triangle A). Find the length x of one of the unknown sides.

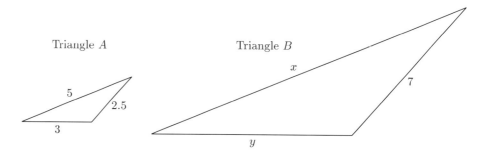

Figure 8

Solution

The relationship between the lengths in the two triangles can be summarized in a table as follows:

Triangle A	Triangle B
2.5	7
5	x
3	y

Notice how the scale factor is found: the length of the side in shape B is divided by the length of the corresponding side in shape A.

The scale factor between the two shapes takes a side of length 2.5 to a side of length 7, so the scale factor must be $7 \div 2.5 = 2.8$. The side of length 5 in triangle A must be enlarged by the same scale factor to become the unknown side in triangle B; therefore the unknown side will be of length $x = 5 \times 2.8 = 14$. This method produces the result by working *between* the shapes (that is, horizontally in the above table).

The unknown length can also be found by looking at the ratios of the sides *within* each shape (that is, working vertically in the above table). For triangle A, the ratio of two of the sides is 2.5 to 5; that is, the second side is twice the length of the first. The same ratio must hold in triangle B, so the unknown side has length $x = 2 \times 7 = 14$.

In this example the calculation was easier when using the ratios *within* the shapes, but that will not always be the case. In each method you have to divide one length by another to find either the scale factor or the ratio. Depending on the numbers involved, either method might be the simpler one (or they might be equally difficult!).

Activity 3

Find the length y of the third side in triangle B in Figure 8.

To assist in describing the features of similar shapes, there is a frequently used notation, which is illustrated in Figure 9.

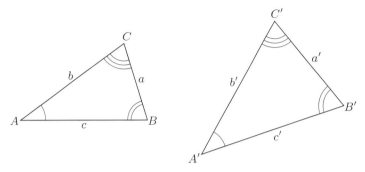

Figure 9   Labelling similar triangles.

Look at the left-hand triangle in Figure 9. The vertices of the triangle are labelled A, B and C. The sides of the triangle can be referred to as AB, BC, and so on. A labelling convention that is shorter and often more convenient involves denoting the sides of a triangle by small letters—in this case, by a, b and c. The side labelled a is always opposite vertex A, the side b opposite the vertex B, the side c opposite the vertex C.

The right-hand triangle in Figure 9 is a scaled version of the triangle on the left. To make it easier to see how the triangles correspond to one another, they have been labelled in a particular way. The vertex A' of the second triangle corresponds to the vertex A of the first. Likewise, vertices B' and C' correspond to the vertices B and C, respectively, and the sides a', b' and c' correspond to the sides a, b and c, respectively. The dash ($'$) notation is a standard way of marking corresponding features. This notation is especially useful when expressing the properties of similar shapes algebraically. Also note that angles marked in the same way are the same size.

Now return to the two relationships concerning lengths in similar shapes (see page 13 and Example 1). Recall that for a pair of similar triangles, these are:

- the scale-factor relationship *between* the triangles; that is,

$$\begin{array}{c}\text{length of a side}\\\text{of one triangle}\end{array} = \begin{array}{c}\text{constant}\\\text{scale factor}\end{array} \times \begin{array}{c}\text{length of corresponding}\\\text{side of other triangle}\end{array}$$

- the relationship between the sides *within* each triangle; that is, the ratio of a pair of sides in one triangle is the same as the ratio of the corresponding sides in the other.

It has been assumed that if one of these relationships is true, so is the other. This can now be justified by using algebra.

Suppose the sides of the left-hand triangle in Figure 9 are each multiplied by the same scale factor, k, in order to achieve the triangle on the right, then

$$a' = ka \quad \text{and} \quad b' = kb.$$

Hence, the ratio of sides a' and b' is

$$\frac{a'}{b'} = \frac{ka}{kb} = \frac{a}{b}.$$

This equation states that the ratio of the sides a and b in the left-hand triangle is equal to the ratio of the sides a' and b' in the right-hand triangle. Thus, if the same scale factor converts each side in one shape into the corresponding side in the other shape, then the ratio between two sides in one shape will be the same as the ratio between the two corresponding sides in the other.

This result is true for *any* pair of corresponding sides.

1.2 *Evidence for establishing similarity*

You have seen how to enlarge a shape to create a similar shape. But, if you are *given* two shapes, how can you tell whether or not they are similar?

It is often useful to establish whether shapes are similar, because that property can be used to find unknown lengths. So, what are the relevant criteria? If you know that the sides of one shape are scaled up by the same scale factor to give the corresponding sides of the other shape and also that the corresponding angles are equal, then the shapes must be similar. However, one of these properties is not enough. For example, if the only thing you know about two shapes is that the sides are all enlarged by the same scale factor, that does not mean the shapes are similar. To see this, look at the shapes in Figure 10. The lengths of the sides of the *parallelogram* are each twice as long as those of the corresponding sides in the *rectangle*, but, of course, they are not similar shapes.

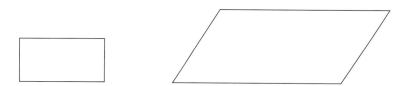

Figure 10   A rectangle and a parallelogram.

Nor is it enough to know just that the corresponding angles are the same sizes. A square and a long thin rectangle both have angles of the same size, but they are not similar shapes.

Triangles are a special case because when the lengths of the three sides of a triangle are known, the angles are fixed; they are unlike shapes with four or more sides in this respect. Thus a triangular framework is rigid, whereas a framework made of four or more rods can be deformed since the angles are not fixed. This property of triangles means that scaling up all the sides of a triangle by the same scale factor *will* always produce a similar triangle. Figure 11 illustrates this.

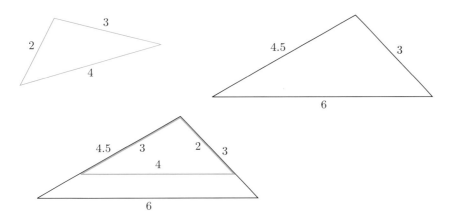

The small triangle needs to be turned over to fit as shown.

Figure 11   Similar triangles.

Because of this special property of triangles, if you want to show that two triangles are similar, it is not necessary to know that each pair of corresponding angles is equal and that the ratios of *all* of the sides within one triangle are the same as the ratios of the corresponding sides within the other triangle. You can show they are similar with more limited information.

To establish that two triangles are similar, there are three frequently used criteria, each depending on only some of the facts about the triangles.

The *first* criterion concerns the lengths of the sides:

- The ratios of the lengths of the sides within one triangle are the same as the ratios of the lengths of the corresponding sides within the other triangle.

This is simply another way of stating the fact that one triangle is a scaled-up version of the other. Knowing the scale factor of enlargement for a pair of triangles is equivalent to knowing that the ratios of the lengths of the three sides are the same in both triangles. This can be illustrated by the triangles in Figure 11. The lengths of the corresponding sides in these triangles are as follows:

This result was proved using algebra in Section 1.1.

Small triangle	Large triangle
2	3
3	4.5
4	6

The scale factor may be calculated from a horizontal comparison in the table: it is 1.5. A vertical comparison in the table shows that in the small triangle, the ratio of the longest side to the shortest side is $4 \div 2 = 2$, and in the large triangle, the corresponding ratio is $6 \div 3 = 2$. The ratio of the intermediate-sized side to the longest side is 0.75 in both triangles. If you know that the ratios of the lengths of the three sides of one triangle are the same as the ratios of the lengths of the three sides of the other, then the triangles must be similar.

The other two main criteria for establishing whether two triangles are similar apply when at least one angle is known to be the same in both triangles. Figure 12 will be useful in what follows. It shows a triangle, ABC, being enlarged with the centre of enlargement at A. You may find it helpful to imagine that triangle ABC is expanding, with the vertex at A remaining fixed while the vertices at B and C move out in such a way that the third side of the triangle always remains parallel to the side BC. The expanding triangle produces a family of similar triangles, some of which are shown in Figure 12.

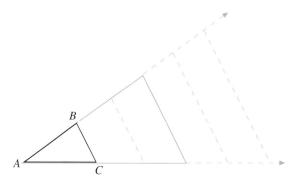

Figure 12   Part of a family of similar triangles produced by enlarging triangle ABC.

Any triangle that is similar to triangle ABC must be able to fit onto one of the family of triangles obtained by enlarging triangle ABC itself. Consider a triangle with one angle equal to angle A. It can be fitted so that two of its sides lie along the lines formed by AB and AC. To show that this triangle is similar to triangle ABC, its third side must be parallel to BC.

One condition that will ensure the third side is parallel to BC is the following: one of the angles of the second triangle must be the same size as angle CBA, and the other angle must be the same size as angle BCA. That is, the three angles of one triangle must each be equal to the three angles of the other. In fact, it is not necessary to know that all three angles are equal: if two of the angles are the same in both triangles, then the third one must also be the same.

The angle sum of any triangle is 180°, so in each triangle the third angle will be 180° minus the sum of the other two angles.

Therefore the *second* criterion for two triangles to be similar is:

- When two angles in one triangle are equal to two angles in the other triangle, the triangles are similar.

An angle contained between two sides is often called an *included* angle.

The third criterion again concerns two triangles with an equal angle, but also involves the ratios of the sides containing that angle. If a triangle is an enlargement of triangle ABC, then it must fit onto one of the family of triangles represented in Figure 12. This will be possible if the two sides containing the angle A are enlargements of AB and AC (by the same scale factor). Put another way, the ratio of those two sides must be the same as the ratio of AB to AC. So the *third* criterion is:

Recall that when the ratios of the lengths of corresponding sides are the same, one triangle will be an enlargement of the other.

- If two triangles have one identical angle, and the sides containing that angle are in the same ratio in both triangles, then the triangles will be similar.

If two given triangles satisfy any one of the three criteria set out above, then you can be confident that they are similar. Figure 13 shows examples of pairs of triangles that satisfy each of the criteria. In Figure 13(a), the two triangles have two angles the same. In Figure 13(b), the ratios of the three sides of one triangle are the same as the ratios of the three sides of the other. In Figure 13(c), each triangle has an angle of 30°, and the sides containing this angle are in the same ratio in both triangles (the longer side is the shorter side multiplied by 4/3).

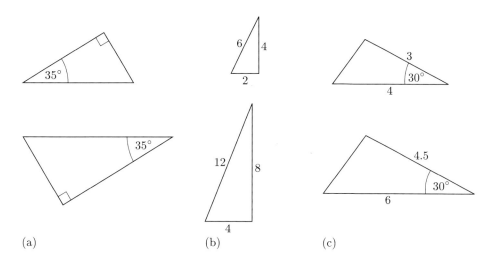

(a)   (b)   (c)

Figure 13   Three pairs of similar triangles.

These three criteria are the ones most commonly used for establishing the similarity of triangles, but two triangles may be similar even when they cannot be compared by using any of these criteria. For instance, an additional criterion applies when both triangles are right-angled:

- If two triangles are right-angled and if the ratio of the pair of sides that does *not* contain the right angle is the same in both triangles, then the triangles are similar.

Of course, if the pairs of sides with equal ratios *do* contain the right angle in both triangles, then the triangles are similar by the third criterion. So a more general criterion is:

- If two triangles are right-angled and if the ratio of *any* pair of corresponding sides is the same, then the triangles are similar.

A word of warning: do remember that these similarity criteria apply only to triangles. For shapes with four or more sides, knowing that all the corresponding angles are equal does not guarantee that the shapes are similar.

The fact that you can tell whether two triangles are similar by comparing just their angles is one reason why triangles play such an important role in geometry. In particular, many shapes can be split into triangles, so it is possible to deduce that other shapes are similar by showing that the triangles forming them are similar.

Activity 4 *Similarity*

Consider the pairs of triangles in Figure 14. Identify those pairs in which the two triangles are similar. (They have not been drawn accurately to scale, so this is not a test of your powers of pattern recognition, but of your ability to use the information given in the diagram.)

Some of the angles are marked with single or double arcs. In each part of Figure 14, features that are marked in the same way are the same size.

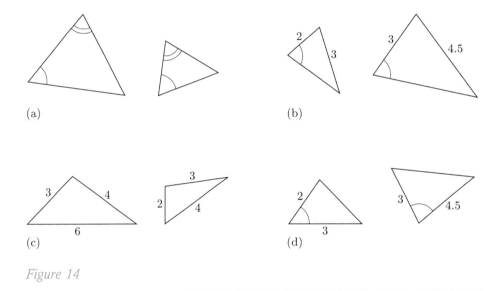

Figure 14

1.3 *Using scale factors and similarity*

The fact that triangles with two matching angles are similar can be used to find unknown lengths, especially lengths that cannot be measured directly, as in the example below.

▶ Figure 15 shows a tree whose height is a matter of dispute. How can its height be measured?

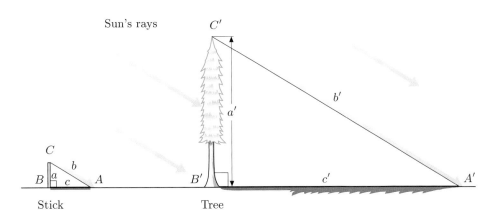

Figure 15   Finding the height of a tree.

Provided that the day is sunny, the tree's height can be found as follows: measure the length of the tree's shadow, and then compare it with the length of the shadow of a stick that is stuck upright in the ground and whose height can be measured easily. The Sun's rays make the same angle with the horizontal ground whether they form the shadow of the tree or the shadow of the stick. If both the stick and the tree are assumed to be vertical, then two pairs of corresponding angles in the triangles ABC and $A'B'C'$ match. The angles at A and A' are both formed by the Sun's rays impinging on the horizontal ground, and so they are the same; also, the angles at B and B' are the angles between a vertical and the horizontal ground hence they are both right angles. Triangles ABC and $A'B'C'$ are, therefore, similar to each other, and so

The Sun's rays are represented by parallel straight lines.

$$\frac{a'}{a} = \frac{c'}{c}.$$

Here a' is the height of the tree, which is unknown. All the other lengths in the equation can be measured directly: a is the height of the stick, c' is the length of the tree's shadow, and c is the length of the shadow of the stick. Rearrangement of the above equation, by multiplying through by a, gives the height of the tree as

$$a' = \frac{ac'}{c}.$$

21

If the height of the stick, a, is measured as 1.5 m, the length of its shadow, c, is 2.5 m and the length of the tree's shadow, c', is 14.5 m, then

$$a' = \frac{ac'}{c} = \frac{1.5 \times 14.5}{2.5} = 8.7\,\text{m}.$$

Therefore the tree is about 8.7 m high. However, this result should be interpreted in the light of the modelling assumptions that have been made.

Activity 5   *Tree modelling assumptions*

Reread the text on page 21 from 'Provided that the day is sunny, ...', and identify any explicit or implicit modelling assumptions that underlie the process of calculating the height of the tree. How might they effect the interpretation of the above result?

Activity 6   *Another tree*

The same process was repeated for a different tree. The height of the stick was 1.2 m, its shadow measured 0.8 m and the tree's shadow measured 6.4 m. Calculate the height of the tree.

Activity 7   *Elevatory, my dear Watson!*

An interesting literary excursion into the ideas of this section is provided by Sir Arthur Conan Doyle in his Sherlock Holmes' story 'The Musgrave Ritual'. Read the extract from that story in the Reader now, and jot down the mathematical techniques referred to in the extract.

1.4 *Proving Pythagoras' theorem*

You may have known Pythagoras' theorem for some time, but do you know how to show it is true for all right-angled triangles?

You could draw a large number of right-angled triangles and check that Pythagoras' theorem applies to all of them, but that would not prove the theorem is true for *all* right-angled triangles. What is needed is a general proof that applies to any right-angled triangle.

Over the centuries since Pythagoras' theorem was formulated, very many proofs have been devised. One of these uses the properties of similar triangles. It starts with a general right-angled triangle, ABC, with the right angle at C. Without specifying the lengths of the sides a, b and c, the proof shows that

$$c^2 = a^2 + b^2.$$

The proof is set out below in several stages, with some essential steps taking the form of activities. Both geometry and algebra are used.

Look at the right-angled triangle ABC on the left of Figure 16. The size of the angle at A is denoted by α, the size of the angle at B by β, and the angle at C is a right angle; the lengths of the sides are denoted by a, b and c. A line CD has been drawn perpendicular to the line AB, dividing triangle ABC into two smaller triangles. The length of AD is denoted by x, and that of BD by y. Note that the length of AB is c, but it is also the sum of AD and DB, so $c = x + y$. The proof uses the fact that triangle ABC and the two smaller component triangles are all similar. For convenience, Figure 16 shows the two smaller triangles placed in corresponding positions to triangle ABC.

Note that the smaller triangles have been turned and then flipped over.

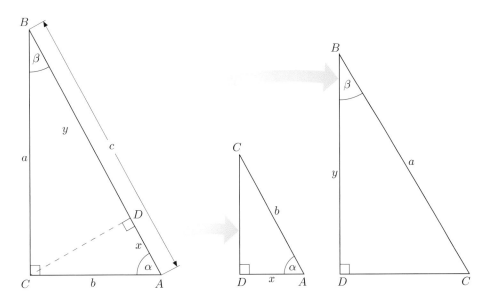

Figure 16   Similar right-angled triangles.

Activity 8   *Similar right-angled triangles*

(a) Explain why the triangle ACD in Figure 16 is similar to the triangle ABC.

(b) Because the triangles are similar, the ratio of a pair of sides in triangle ACD will be the same as the ratio of the corresponding sides in the triangle ABC. The ratio of two of the sides in triangle ACD is x/b. What is the corresponding ratio in triangle ABC?

Since the triangles are similar, the two ratios are equal. This means that

$$\frac{x}{b} = \frac{b}{c}.$$

Multiplying through by b gives

$$x = \frac{b^2}{c}.$$

The third triangle in Figure 16, CBD, is also similar to triangle ABC. This is because there is an identical pair of angles in the two triangles: both triangles have a right angle and an angle β. The same method as that just employed to obtain an expression for x can be used in these two triangles to obtain an expression for y. The expressions for x and y can then be substituted into the equation $c = x + y$, giving an equation that contains just a, b and c.

Activity 9   *More similar right-angled triangles*

Because the triangles CBD and ABC are similar, the ratio of a pair of sides in one of these triangles will be the same as the ratio of the corresponding sides in the other triangle. The ratio of two of the sides in triangle CBD is y/a. Use the corresponding ratio in triangle ABC to show that

$$y = \frac{a^2}{c}.$$

Now substitute for x and y in the equation $c = x + y$, and multiply through the resulting equation by c to show that $c^2 = a^2 + b^2$.

In this proof you have used the properties of similar triangles to establish that for any right-angled triangle ABC (with the right angle at C), $a^2 + b^2 = c^2$. Since the proof did not depend upon specific values of the lengths a, b and c, or the specific sizes of the angles α and β, the proof is general and applies to all right-angled triangles. Hence Pythagoras' theorem applies to all right-angled triangles.

1.5 Similarity, circles, angles and arcs

It is important to appreciate that any two circles are similar. Thus, all lengths in one circle will have to be multiplied by the same scale factor to give the corresponding lengths in the other.

Consider a circle whose diameter is of length 1 unit. The length of the circumference of such a circle will be π. A second circle whose diameter is d and whose circumference is c will be similar to the first (see Figure 17).

π is often defined as the circumference of a unit circle, as here.

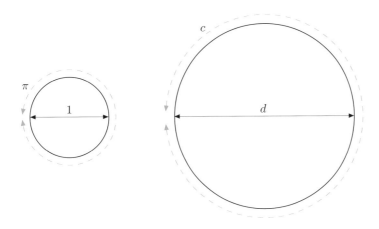

Figure 17   Two similar circles.

The lengths in these two circles are compared in the table below.

	Circle 1	Circle 2
Diameter	1	d
Circumference	π	c

Enlarging circle 1 into circle 2 must change a diameter of length 1 into a diameter of length d, and so the scale factor must be d. Therefore, the circumference of the second circle must be

$$c = \pi \times d.$$

This is the formula for the circumference of a circle of diameter d.

Since $d = 2r$, where r is the *radius* of the circle,

$$c = 2\pi r.$$

The properties of similar circles can also be used to find the lengths of *arcs* of circles. The length of an arc can be specified by the angle at the centre of the circle, as in Figure 18 overleaf.

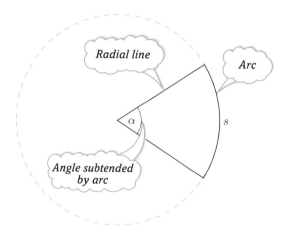

Figure 18   An arc and subtended angle.

The angle at the centre of the circle is often called the *angle subtended by the arc*. Alternatively, you can say that *the arc subtends the angle*.

Recall from *Unit 9* that angles can be measured in either degrees or radians.

Arc lengths are most easily found when angles are measured in radians. Radians are in some sense a more natural way of measuring angles than degrees: they do not use an arbitrary unit like degrees. This is illustrated in Figure 19, where the circle is of radius 1 unit, and the number line, which is a tangent to the circle, uses the same units.

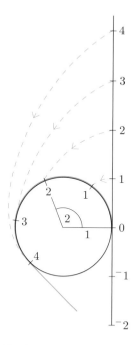

Figure 19   Wrapping the number line.

The number line is 'wrapped around' the circle so that each point on the line falls on the circumference of the circle. An angle is measured by the number of the corresponding point on the wrapped number line. This number is the size of the angle in radians. Hence the size of each angle in radians is the length of the arc defined by the angle. For example, an angle of 2 radians is subtended by an arc length of 2 (see Figure 19), and an angle of 0.5 radian is subtended by an arc length of 0.5. To generalize: an angle of θ radians is subtended by an arc length of θ. A half-turn, an angle of 180°, will have a radian measure equal to half of the circumference. As the radius of the circle is 1, half the circumference is π. So an angle of 180° has a value of π when measured in radians. Similarly, a full turn, 360°, has a value of 2π.

Recall that θ is the Greek (lowercase) letter 'theta'.

This provides an easy method of finding an arc length in a circle of any radius. Since all circles are similar, any circle is a scaled-up version of the circle of unit radius, as shown in Figure 20. So a circle of radius r will have a scale factor of r, and in that circle the arc length l corresponding to an angle θ will be $r\theta$. Therefore

$$l = r\theta.$$

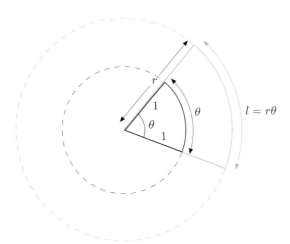

Figure 20   Arc lengths for circles of radius 1 and radius r.

It follows that, for example, when the radius is 3 m, an angle of 2 radians will have an arc length of $3 \times 2 = 6$ m, and an angle of 0.5 radian will have an arc length of $3 \times 0.5 = 1.5$ m.

Degrees and radians each have advantages and disadvantages as the units of measurement for angles: radians are more useful when calculating arc lengths of circles, whereas degrees are often preferable for measuring bearings. Degrees are favoured for bearings because a full turn is a whole number of degrees, as are many useful fractions of a turn: for instance, a half turn, a third of a turn, and a quarter turn.

Radians are very important in more advanced mathematics.

Remember that if there is no degree sign (°), then, by convention, the angle measurement is in radians.

As already pointed out, an angle of $180°$ is π radians, and so

$$1 \text{ radian} = \frac{180°}{\pi} \simeq 57°, \text{ and } 1° = \frac{\pi}{180} \text{ (radian)} = 0.0175 \text{ (radian)}.$$

Figure 21 shows the comparative sizes of $1°$ and 1 radian.

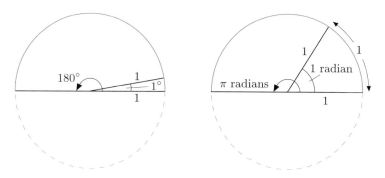

Figure 21 $1°$ and 1 radian compared.

The formula for the arc of a circle is more complicated if the angle is in degrees, and consequently it is often simpler to convert to radians and then use the formula $l = r\theta$.

A similar formula was known to the Greeks around 250 BC, and was used by Eratosthenes to work out a value for the circumference of the Earth. It is said that, on a particular day, Eratosthenes was at a place called Syene in Egypt (modern Aswan), which he knew to be some 800 km due south of Alexandria. He observed that the Sun at midday shone directly down the shaft of a deep well, and the sunlight was reflected straight back into his eyes as he looked into the well. The Sun was, therefore, directly overhead at Syene at noon on that day. Eratosthenes found out that at Alexandria on the same day the direction of the Sun at noon was $7.2°$ south of the vertical. Assuming the Sun's rays to be parallel as they arrive at the Earth's surface, he deduced that an angle of $7.2°$ gives an arc of 800 km on the Earth's surface, as illustrated in Figure 22.

Of course, Eratosthenes did not use kilometres but *stades*: 1 stade $\simeq \frac{1}{10}$ mile. So Eratosthenes' figure of 5000 stades $\simeq$ 500 miles $\simeq$ 800 km. The angle of $7.2°$ is a conversion from Eratosthenes' value of 'one fiftieth of a whole turn'.

To use the formula $l = r\theta$, the angle here should be converted into radians:

$$7.2° = 7.2 \times \frac{\pi}{180} = \frac{\pi}{25} \text{ (radian)}.$$

So with $\theta = \pi/25$, $l = 800$ km, and the radius of the Earth as r km, the formula $l = r\theta$ gives

$$800 = r \times \frac{\pi}{25}.$$

Multiplying through by 25 and dividing by π gives

$$r = \frac{25 \times 800}{\pi}.$$

Hence

$$\text{Earth's circumference (in km)} = 2\pi r = 2\pi \times \frac{25 \times 800}{\pi} = 40\,000.$$

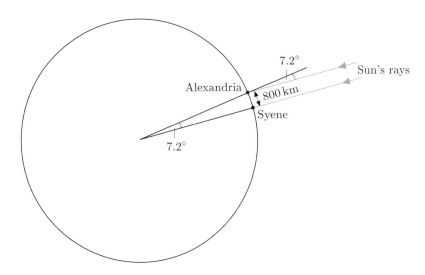

Figure 22   Eratosthenes' method.

The data used in the preceding calculation are not very accurate. The length of an ancient Greek stade is not known exactly, so the distance of 800 km is probably only accurate to one significant figure. Rather surprisingly, the value of 40 000 km is accurate to two significant figures—the polar circumference of the Earth is close to 39 900 km.

The formula $l = r\theta$ was very useful here, and it will be used later to calculate distances between places on the surface of the Earth.

Activity 10 *The Earth's radius*

Calculate the radius of the Earth, using Eratosthenes' data. What modelling assumptions can you identify underlying the calculation?

This section has covered a number of important ideas. In summary, the basic features of scale maps exemplify the mathematical concept of similarity. Two figures are similar when one is an enlarged version of the other: all lengths are scaled by the same scale factor, and all corresponding angles are the same in both figures.

Because many geometric figures can be broken down into a number of triangles, the minimum amount of evidence needed to be sure that two given triangles are similar is well documented.

The idea of similarity is also very useful when applied to circles and circular arcs.

Outcomes

After studying this section, you should be able to:

◇ consider the strategy and mathematical skills needed for the study of the topics in this unit (Activity 1);

◇ decide if two figures (in particular, triangles) are similar (Activities 2 and 4);

◇ use the fact that figures are similar to deduce other information about sides and/or angles (Activities 2, 5 and 6);

◇ prove Pythagoras' theorem using the properties of similar triangles (Activities 8 and 9);

◇ use the formula $l = r\theta$ relating the length of an arc of a circle to the radius and the angle (in radians) subtended by the arc (Activity 10);

◇ identify the mathematical techniques referred to in written accounts (Activity 7).

2   *Getting things into perspective*

Aims   This section aims to elucidate the role of similarity in perspective drawings.   ◇

The properties of similar triangles are useful in the world of art, particularly in representing three-dimensional scenes in two-dimensional perspective pictures. The representation of an object in a picture is, of course, only two-dimensional and it may, therefore, be a very different shape to the original. Thus the shape that represents a rectangular door may not itself be a rectangle.

Figure 23 shows a painting entitled *Self-criticism* by contemporary London-based artist Patrick Hughes, whose work you will see more of in the video band associated with this section. Look at the representation of the doorframe on the left-hand wall and at the representation of the door, which is the same shape, on the other side of the room. The picture draws attention to the difference between what people see an object to be and what they know it must be like. When you look at the picture of the doorframe, you assume that in reality it is a rectangular shape, as would be the door that fits it. However, in the picture, the doorframe is not represented by a rectangle: it is a trapezium (a four-sided figure with one pair of sides parallel). The representation of the doorframe looks right to the viewer, as it is drawn in perspective. On the other hand, the representation of the door, which has been drawn out of place, does not look right, as in this position it is not in perspective.

Figure 23   *Self-criticism* by Patrick Hughes.

Perspective drawing is a technique used in much Western representational painting to ensure that the resulting image 'looks right'. When looking at such a painting, you see the two-dimensional shapes as versions of real three-dimensional objects: you see the trapezium in Patrick Hughes' picture *as* a rectangular doorframe.

Perspective can be explored mathematically. As you will see, the properties of similar triangles play an important role in analysing the processes of constructing perspective pictures and reconstructing the reality that they represent.

2.1   The principles of perspective drawing

Figure 24 shows an artist drawing a scene in *central* perspective—that is, as seen from one fixed point. This fixed point corresponds to the position of the artist's eye, and is known as the *viewpoint*. The *picture plane* is a fixed vertical plane containing the picture. When the picture is seen from the viewpoint, the rays of light entering the viewer's eye should appear just as if they had come from the object itself. In effect, looking at the finished picture is like looking at the scene through a window.

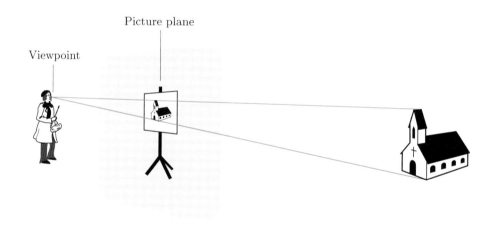

Figure 24   Drawing in perspective.

Imagine you *are* looking at a scene through a window and drawing a perspective painting on that window. You can do this by putting a mark on the glass corresponding to each feature of the scene. Each mark would be made on the glass at the point where a ray of light from the feature passes through the glass to your eye.

It might be practicable to draw a picture in this way on a sheet of glass, but it would be difficult to do this on a canvas, as the canvas itself would obstruct the light. Various ways of carrying out the process in practice were suggested by European artists in the sixteenth and seventeenth centuries, such as the method illustrated in a woodcut by the German artist Albrecht Dürer (see Figure 25).

Figure 25   *Man Drawing a Lute* by Albrecht Dürer.

In this method, a string, representing a ray of light, is stretched from the viewpoint (a pulley on the wall) to a point on the object—in this case, a lute (a type of musical instrument). A wooden frame represents the picture plane, and the point where the string passes through the frame gives the corresponding point on the proposed picture of the object. The method has been described as follows:

> The string from the pulley to the pointer represents a single ray of light and passes through the picture plane. As the man with the pointer fixes different reference points on the lute, his assistant measures off the vertical and horizontal coordinates and plots each new point on the drawing. When there are enough points, he joins the relevant ones, and completes the drawing.
>
> Fred Dubery and John Willats, (1983) *Perspective and Other Drawing Systems*, The Herbert Press, London, pp. 70–1.

It is doubtful whether many perspective paintings were made using this laborious method, but the same principles underlie all perspective paintings. A more manageable method than that just described exploits the mathematics of similar triangles, making it possible to calculate positions in the painting, rather than having to measure them. By knowing the sizes of the real objects and their distances from the artist, the sizes of the objects in the picture can be calculated.

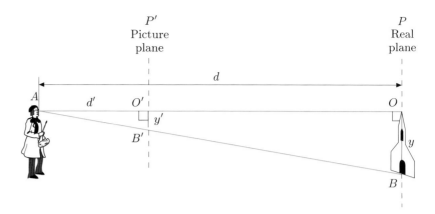

Figure 26   Scaling in perspective.

An idealized model of an artist painting a church is shown in Figure 26. Here A is the artist's viewpoint, P' is the vertical plane containing the canvas (the *picture* plane), and P is a vertical plane in the scene to be painted (the *real* plane). The distance BO represents the height of the actual church steeple, and $B'O'$ is the height of its representation in the picture. Assume that both the steeple and the picture are vertical, and that the angles AOB and $AO'B'$ are both right angles. It then follows that the triangles AOB and $AO'B'$ are similar as they have two corresponding angles equal: the right angles, and the angle at A which is common to both triangles. Since the triangles are similar, one triangle is a scaled version of the other. But what is the scale factor? This factor can be calculated by comparing the lengths in the table below.

	Real object	Picture
Distance	d	d'
Height	y	y'

The scale factor has to be the same for both the distance and the height, and so $d' = kd$ and $y' = ky$, where k is the scale factor. Hence

$$k = \frac{d'}{d}.$$

This gives

$$y' = y \times k = y \times \left(\frac{d'}{d}\right).$$

This equation states that the scale factor for the height of an object in the painting is the distance d' from the artist's viewpoint to the picture plane divided by the distance d from the artist's viewpoint to the actual object. The same is true for the width of an object; indeed, y can represent any dimension in the same plane.

The fact that perspective pictures use the same scale factor for all dimensions in the same plane leads to a small amount of distortion. Objects that are the same distance from the artist's eye are, actually, on the surface of a sphere rather than on a plane. When the sphere (and the objects) are some distance from the artist, a plane is a reasonable approximation, but when the sphere (and the objects) are close to the artist, there will be appreciable distortion (see Figure 27).

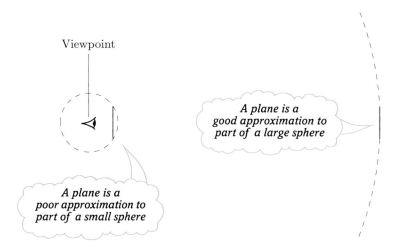

Figure 27   Planes approximating spheres.

For a particular picture of a scene, the distance d' of the picture plane from the artist is fixed, but the distance d of the objects from the artist varies. The bigger the distance d of an object, the smaller the scale factor k. Consequently, the further away an object is, the smaller will be its representation in the picture. This gives rise to the concept of the vanishing point.

Remember that $k = d'/d$ and so k is inversely proportional to d. See *Unit 13*.

To explore this further, consider a set of parallel lines on the ground, receding into the distance directly away from the artist. These might be, for example, floorboards or railway lines. The width of a floorboard or the distance between a pair of railway lines is constant, but the length representing this dimension in a perspective picture will change, as it is inversely proportional to the distance d from the artist. Therefore, in the picture, the representations of the parallel lines get closer as they recede into the distance, and eventually they meet at a point, called the *vanishing point*, as in Figure 28.

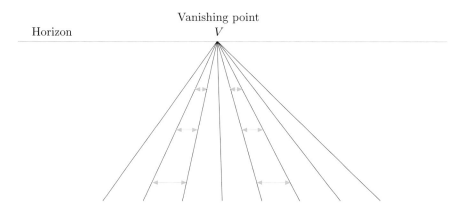

Figure 28   The vanishing point.

In perspective paintings, it is not only parallel lines on the ground that meet at a vanishing point. For example, the horizontal lines on a wall, which are formed by the top of the skirting board, the window ledges and the ceiling edge, also meet at a vanishing point. In general, any set of lines that are parallel to a line on the horizontal will meet at a vanishing point. All the vanishing points lie on a horizontal line called the *horizon*; this corresponds to the actual horizon when the ground is completely flat.

As you will see in Section 2.2, the vanishing point and the horizon can be very useful in constructing lines in a perspective picture, thus avoiding the difficulties of a method like Dürer's.

2.2   Drawing in perspective on your calculator

There is a short section (Section 14.3) in the *Calculator Book* that uses a program to draw simple perspective pictures like that in Figure 28. It shows how changing the relative positions of the viewpoint, the picture plane, and the scene that is being drawn all affect the final picture.

Now read the introduction to Chapter 14 of the Calculator Book, and then work through Section 14.3. Leave Sections 14.1 and 14.2 until later in the unit.

2.3   Vermeer's perspective

One painter who excelled at the portrayal of interior scenes in perspective was the Dutch artist, Jan Vermeer (1632–75). He used the vanishing point to help him in constructing the perspective. Moreover, he exploited the effect of tiled floors to emphasize the perspective (such floors were popular in Holland at the time). However, he often made the construction a little less obvious, by setting the tiles at an angle of 45° to the picture plane. The parallel lines of tiles on the real floor are represented in the painting by lines that meet at a point on the horizon: such points are called *distance points*. This is illustrated in Figure 29.

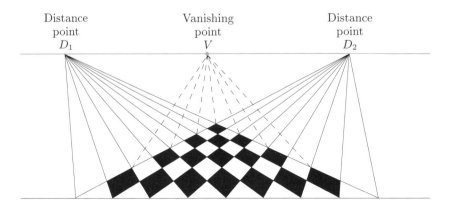

Figure 29   A tiled floor with a vanishing point and distance points.

One of Vermeer's paintings, which is the subject of the video band associated with this section, is reproduced in Figure 30.

Figure 30   *The Music Lesson* by Jan Vermeer.

The perspective in this painting is so accurate that it has proved possible to reconstruct the actual scene from measurements in the painting itself. The relative dimensions of the room and the positions of the instruments, the player and the music master, as well as the position from which Vermeer painted the picture, can be worked out using the principles of projective geometry and similar triangles.

Activity 11   *Vermeer's methods (optional)*

Study the Reader article 'Vermeer in perspective' by Jorgen Wadum, and jot down the key points about perspective drawing.

Now view band 10 of DVD00107 'Getting things into perspective'.

Projective geometry

Projective geometry is a branch of mathematics that takes as its main premise the idea that geometrically similar figures can be superimposed exactly on top of one another from the monocular viewpoint of a single-eyed observer. Figure 31 shows two triangles, *ABC* and *A'B'C'*, that are 'the same' when considered from this viewpoint (represented by the picture of an eye).

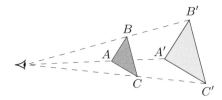

Figure 31   Two triangles in perspective.

Such geometry relies heavily on the notions and properties of similar triangles, but its results are true for all figures that are projectively related in this way.

Activity 12   *Points of view*

Consider the geometry of the room you are in. Focus on an object and move around it, drawing different sketches of it from different points of view. Try to focus on what you actually *see* and not what you 'know' about the object. Repeat this for other objects. Can you explain why different viewpoints produce different sketches?

Look at the way that the walls and ceiling meet at the corners, and at the shapes of the doors and windows. You 'know' that the top and bottom of a door or window are parallel – but do they *look* parallel from your viewpoint? (Recall Patrick Hughes' painting in Figure 23.) Can you explain what you see?

Activity 13   *Understanding Vermeer's methods (optional)*

Study the Reader article 'The photographic accuracy of Vermeer's paintings' by Fred Dubery and John Willats.

Jot down the key points about reconstructing the room from Vermeer's perspective painting.

Activity 14   *Explaining perspective*

Make notes on the principles of perspective drawing, using diagrams as well as words if you wish. Include an explanation of why lines of equal length in a drawing do not always represent lines of equal length in reality.

If you completed Activity 11, add these notes to the ones you produced there.

Geometric language in everyday use

There are a number of examples of geometric ideas and images being used in everyday English to refer to intellectual, cultural or social matters. Speakers take 'positions', they offer 'perspectives', illustrating 'points of view' and giving 'views' (meaning 'opinions'), following 'lines' of argument, and sometimes arguing 'circularly'.

Finally, there is one more level on which to think about all this. You probably watched the video band on a (slightly curved) two-dimensional screen. How aware were you when watching that all the images you were experiencing were flat? How did it affect the way in which you interpreted what you saw as two- or three-dimensional?

Outcomes

After studying this section, you should be able to:

◇ understand the application of similar triangles to perspective drawing, with particular reference to the paintings of Vermeer (Activities 11, 12, 13 and 14).

◇ be alert to some of the problems in interpreting two-dimensional representations of three-dimensional objects (Activities 11 and 12).

3 *Similarity and trigonometry*

Aims   This section aims to build upon the ideas about similarity that were introduced in Section 1 and to link them to the trigonometric functions that you met in *Unit 9*.   ◇

Trigonometric functions can be applied to triangles by using the ideas of similarity. The methods developed in this section are widely employed in mathematics and its applications. They rely on the fact that similar figures are scaled versions of each other and, consequently, information about one figure (usually a 'standard' figure) can be used to find corresponding information about others. The standard figure is customarily chosen to be as simple as possible; for example, the standard circle is chosen to have a radius of 1 unit.

3.1 *Similarity and sines*

In Section 1.3 of *Unit 9*, you met the sine function. It was defined in terms of a circle of radius 1 unit—the unit circle.

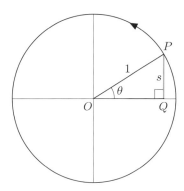

Figure 32   The unit circle.

Several later diagrams are based on the *unit circle*.

Recall that, as a point P moves anti-clockwise round the unit circle, the angle θ increases in size (see Figure 32). The vertical height of P, above the horizontal axis, is s. This height gives the sine of the angle θ. Thus $s = \sin \theta$.

The value of s, and correspondingly of $\sin \theta$, increases from 0 when θ is 0, to 1 when θ is $\pi/2$, or 90°, and subsequently decreases back to 0 when θ is π, or 180°. Then, as θ increases beyond π, P moves below the horizontal axis. This means that s, and hence $\sin \theta$, become negative, dropping to $^-1$ when θ is $3\pi/2$, or 270°, and finally returning to 0 when θ is 2π, or 360°, to complete one cycle.

Now consider Figure 33, which shows an enlargement of the unit circle, together with the original unit circle from Figure 32, both drawn with the same centre O. The scale factor of enlargement is r, therefore all the lengths in the enlarged circle will be equal to the corresponding lengths in the unit circle but multiplied by r: for instance, the enlarged circle will have a radius r. However, the angles will be unchanged.

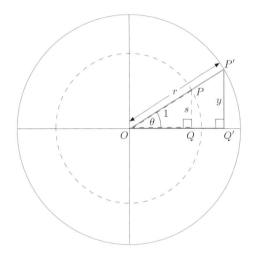

Figure 33   An enlargement of the unit circle.

Several later diagrams are based on an enlargement of the unit circle.

The scaling involved in going from the unit circle to the enlargement is summarized in the following table:

	Unit circle	Enlargement of unit circle
Radius	1	r
Height	s	y

As the scale factor is r, it follows that $y = r \times s = r \sin \theta$. Dividing through by r gives

$$\sin \theta = \frac{y}{r}.$$

This is true for a circle of any size, where r is the radius and y is the height, above the horizontal, of P'—the point at which the radius drawn at an angle θ to the horizontal meets the circumference.

Although there is a sine function for every angle whatever its size, many applications involve right-angled triangles and so only concern angles between 0 and $\pi/2$, or 90°. In such cases you need only consider P' in the first quadrant, and the angle θ will be acute.

The right-angled triangle $OP'Q'$ in Figure 33 is reprinted overleaf without the circle, as Figure 34. The side $P'Q'$, which has length y, is the side *opposite* the angle, and the side OP', which has length r, is the *hypotenuse*. For right-angled triangles, this leads to an alternative definition of sine in terms of the sides of the triangle:

$$\sin\theta = \frac{y}{r} = \frac{\text{opposite}}{\text{hypotenuse}}.$$

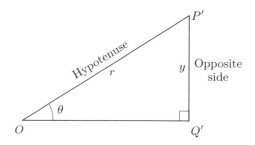

Figure 34

The sine function provides a means of finding lengths in a right-angled triangle. If you know the angle θ and either of the lengths y or r, you can find the other one of these lengths from $\sin\theta$.

The values of the sines are conveniently available at the press of a key on your calculator. However, if you are old enough to remember life before calculators, you may recall having to look up sines in a book of tables—the values of the sines of different angles have been tabulated since the fifth century (in a Hindu work).

> The calculator does not actually store the values of sine for all possible angles. It stores a number of values and then uses formulas to calculate the values for other angles from them.

Example 2   *What is y?*

Look at Figure 35. Given that OP' is 1.5 m long and the value of $\sin 60°$ is 0.866, what is the length of the side $P'Q'$?

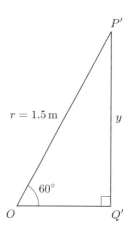

Figure 35

Solution

Side $P'Q'$ has length y metres. Now,

$$\sin 60° = \frac{\text{opposite}}{\text{hypotenuse}} = \frac{y}{r}.$$

Multiplying through by r gives

$$y = r \times \sin 60° = 1.5 \times 0.866 = 1.299 = 1.30 \text{ (to 3 significant figures)}.$$

So $P'Q'$ has length $1.30\,\text{m}$.

Example 3   *Find r*

In Figure 34, suppose the angle θ is $\pi/6$ and the length y is $2\,\text{cm}$. Find the length, r cm, of the hypotenuse.

Solution

By definition $\sin\theta = \dfrac{y}{r}$. Substituting in the given values of θ and y, you obtain $\sin\dfrac{\pi}{6} = \dfrac{2}{r}$. From your calculator, $\sin\dfrac{\pi}{6} = 0.5$. Therefore

$$\sin\frac{\pi}{6} = 0.5 = \frac{2}{r}.$$

In the equation

$$0.5 = \frac{2}{r},$$

multiplying through by r to get rid of the fraction gives $0.5r = 2$.

Hence

$$r = \frac{2}{0.5} = 4.$$

Thus the length of the hypotenuse is $4\,\text{cm}$.

You might like to refer back to Chapter 9 of the *Calculator Book* to remind yourself how to find the sine function (and its inverse) on the course calculator. Remember to have your calculator in radian mode.

Activity 15   *Find y and r*

(a) In a right-angled triangle, the hypotenuse has length $10\,\text{cm}$ and one angle is $\pi/6$. What is the length of the side opposite this angle?

(b) In another right-angled triangle, one angle is $\pi/4$ and the length of the side opposite it is $8\,\text{cm}$. Find the length of the hypotenuse to 3 significant figures.

If, instead of knowing the values of an angle and a length, you know the values of the two lengths y and r and want to find the angle θ, then you need the inverse sine function, which you met in Chapter 9 of the

Calculator Book. The *inverse function* of the sine is often denoted by $\sin^{-1}$, but you may also see it referred to as arcsine, arcsin or, occasionally, asin.

Example 4   *What is θ?*

A proposed layout for a herb garden is shown in Figure 36. The length $AB = y = 4\,\text{m}$ and the length $OB = r = 5\,\text{m}$. Find the angle θ by using the inverse sine function.

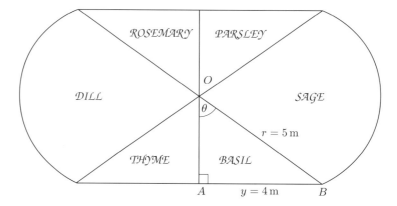

Figure 36   A herb garden.

Solution

In the triangle, AOB, designated for basil,

$$\sin\theta = \frac{y}{r} = \frac{4}{5} = 0.8.$$

Here you know the sine of the angle, but you want to know the angle itself. The inverse sine function undoes the sine function, so $\sin^{-1}(0.8)$ will give you the angle θ whose sine is 0.8. The calculator gives the value of θ as just over $53°$ or about 0.927 radian.

The calculator gives the acute angle whose sine is 0.8, but there is also an obtuse angle in the second quadrant, whose sine is 0.8 (it is about $127°$). However, here it is the acute angle that is required. There is further discussion of this in Section 3.6.

Activity 16   *Find θ*

(a) In a right-angled triangle, the length of the hypotenuse is $10\,\text{cm}$ and the length of one of the other sides is $6\,\text{cm}$. What is the size of the angle opposite this side? Give your answer correct to the nearest degree.

(b) In another right-angled triangle, the length of the hypotenuse is $8\,\text{cm}$ and that of one of the other sides is $5\,\text{cm}$. What is the size (in radians) of the angle opposite this side? Give your answer correct to 2 significant figures.

3.2 *The cosine and tangent functions*

In Chapter 9 of the *Calculator Book* you met two other trigonometric functions, namely *cos* (short for *cosine*) and *tan* (short for *tangent*). Like the sine, these functions can be defined in terms of a point P moving round the unit circle. As you have already seen, the sine of an angle is equal to the vertical height, s, of the point P in the unit circle. Similarly, the *cosine* of the angle θ is equal to the corresponding horizontal distance, c, in the unit circle (see Figure 37).

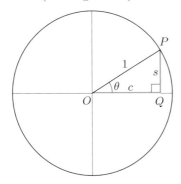

Figure 37   The unit circle.

Consider what happens to c, and hence to $\cos\theta$, as the point P travels anti-clockwise round the circle and the angle θ increases. Initially, when θ is 0, P and Q will coincide on the circumference of the unit circle, and c ($= \cos\theta$) will be 1. Then, as P moves round the circle, and θ increases, the distance c, and correspondingly $\cos\theta$, decrease, becoming 0 when θ is $\pi/2$, or 90°. As P moves into the second quadrant, the value of c, and of $\cos\theta$, becomes negative, reaching a minimum of $^-1$ when θ is π, or 180°. Thereafter, c, and $\cos\theta$, increase again, first to 0 when θ is $3\pi/2$, or 270°, and back to 1 when θ is 2π, or 360°, at the end of a complete turn.

As in the case of the sine function, Figure 37 can be enlarged to produce Figure 38.

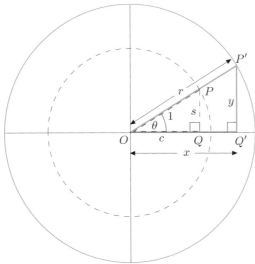

This is the same enlargement with scale factor r as that used in Section 3.1.

Figure 38   An enlargement of the unit circle.

Activity 17 *Check cosine*

Use the fact that the triangles OPQ and $OP'Q'$ in Figure 38 are similar to show that

$$\cos\theta = \frac{x}{r}.$$

Hint: use an analagous argument to that used on page 41 for $\sin\theta$.

Like the sine function, the cosine function is often used for angles in right-angled triangles. For example, consider the triangle $OP'Q'$ in Figure 39 (taken from the enlargement of the unit circle in Figure 38). Within this triangle, the side OQ', which has length x, is the side *adjacent* to the angle θ, and the side OP', which has length r, is the *hypotenuse*.

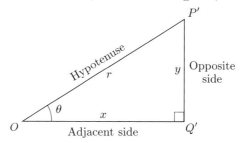

Figure 39

Thus, for right-angled triangles, the cosine function can be defined as

$$\cos\theta = \frac{x}{r} = \frac{\text{adjacent}}{\text{hypotenuse}}.$$

Example 5 *What is x?*

In Figure 40, the angle at A is $60°$ and the length of AB is $3.5\,\text{m}$. Find the length of the side AC.

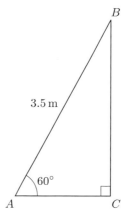

Figure 40

Solution

Let the length of AC be x metres. Then

$$\cos 60° = \frac{\text{adjacent}}{\text{hypotenuse}} = \frac{AC}{AB} = \frac{x}{3.5}.$$

Multiply through by 3.5 to obtain

$$3.5 \times \cos 60° = x.$$

The calculator gives $\cos 60°$ as 0.5. So

$$x = 3.5 \times 0.5.$$

Therefore, the length of AC is 1.75 m.

A tangent to a circle is a line that just touches the circle. The tangent function of an angle θ is related to this. In the unit circle in Figure 41, the line AB is a tangent line. The radius OP, when extended, meets this line at B. The tangent of the angle θ is given by the length, t, of AB. Thus $t = \tan\theta$.

The Latin verb *tangere* means 'to touch', as in tangible.

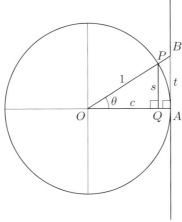

Figure 41   The unit circle with tangent.

Think once more about the point P travelling anti-clockwise round the unit circle. As the angle θ increases from 0 to $\pi/2$, or 90°, the length t, and correspondingly $\tan\theta$, get larger. But at $\pi/2$, OP becomes parallel to the tangent AB, so t, and hence $\tan\theta$, cannot be defined here.

As P moves still further round the circle, the line OP can be extended *backwards* to intersect the tangent AB. The distance, t, of the intersection from A again gives the value of $\tan\theta$. Because this distance is below A, $\tan\theta$ has a negative value. When P reaches the point where the angle θ is π, or 180°, the extended line will pass through A, and so $\tan\theta$ will be 0. As P travels further on, $\tan\theta$ takes positive values until θ is $3\pi/2$, or 270°, when it is again undefined. Thereafter, P moves into the fourth quadrant, and $\tan\theta$ becomes negative. Finally, P returns to the start when θ is 2π, or 360°, and $\tan\theta$ is again 0.

Note that in the second and third quadrants, OP is extended backwards to meet AB, while in the first and fourth quadrants, OP is extended forwards.

In Figure 41 the triangles OPQ and OBA are similar (two corresponding angles match, namely the right angles and θ). This leads to an important relationship between s, c and t, and therefore between the sine, cosine and tangent functions. The corresponding lengths in the triangles in Figure 41 are given in the following table:

Triangle OBA	Triangle OPQ
$OA = 1$	$OQ = c$
$AB = t$	$PQ = s$

The scale factor for this enlargement takes $OA = 1$ to $OQ = c$, and so must be c. The same scale factor takes t to s, and so $s = t \times c$.

To make t the subject, divide this equation through by c and obtain

$$t = \frac{s}{c}.$$

Since $s = \sin\theta$, $c = \cos\theta$ and $t = \tan\theta$, written in full $t = \dfrac{s}{c}$ becomes

$$\tan\theta = \frac{\sin\theta}{\cos\theta}. \tag{1}$$

Recall that in the enlargement of the unit circle (see Figure 42), $\sin\theta = \dfrac{y}{r}$ and $\cos\theta = \dfrac{x}{r}$. Substituting for $\sin\theta$ and $\cos\theta$ in expression (1) gives $\tan\theta$ in terms of x, y and r:

$$\tan\theta = \frac{\sin\theta}{\cos\theta} = \frac{y}{r} \div \frac{x}{r}$$

$$= \frac{y}{r} \times \frac{r}{x} = \frac{y}{x}.$$

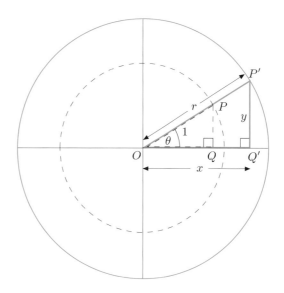

Figure 42   An enlargement of the unit circle.

The tangent of an angle θ can, therefore, be defined as

$$\tan \theta = \frac{y}{x}.$$

Like the sine and cosine functions, the tangent function is often used in right-angled triangles. So, for many applications you only need to consider angles in the first quadrant. Take, for instance, the right-angled triangle $OP'Q'$ in Figure 42. Within this triangle, the side $P'Q'$, which has length y, is the side *opposite* the angle θ, and the side OQ', which has length x, is the side *adjacent* to the angle θ. Then

$$\tan \theta = \frac{y}{x} = \frac{\text{opposite}}{\text{adjacent}}.$$

Thus, in a right-angled triangle, the tangent, like the sine and the cosine, is usually specified as a ratio of lengths. The ratio forms of sine, cosine and tangent are called *trigonometric ratios*, and the study of trigonometric ratios is called *trigonometry*.

The trigonometric ratios are very useful for finding unknown lengths and angles in right-angled triangles when you know some of the lengths and angles.

Trigon is an alternative name for a triangle: 'tri' means 'three', and 'gon' comes from the Greek for 'angle'. So the word 'trigono-metry' can be seen to refer to the measurement of triangles.

Example 6   *Finding angles and sides*

The right-angled triangle shown in Figure 43 has perpendicular sides of 3 and 4 units.

(a) Use the tangent ratio to find the angles of this triangle.

(b) Use the cosine ratio to find the length of the hypotenuse.

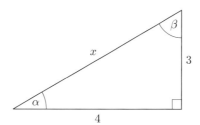

Figure 43

Solution

(a) $\tan \alpha = \dfrac{\text{opposite}}{\text{adjacent}} = \dfrac{3}{4} = 0.75.$

So

$$\alpha = \tan^{-1}(0.75) = 36.86989765°.$$

Then

$$\beta = 90° - \alpha = 53.13010235°.$$

(b) $\cos \alpha = \dfrac{\text{adjacent}}{\text{hypotenuse}} = \dfrac{4}{x}.$

Multiplying through by x and dividing through by $\cos \alpha$ gives

$$x = \frac{4}{\cos \alpha} = \frac{4}{\cos \left(36.86989765°\right)} = 5.$$

So, the hypotenuse has a length of 5 units.

This could also be found using Pythagoras' theorem: $x^2 = 3^2 + 4^2 = 5^2.$

A right-angled triangle with sides of 3, 4 and 5 units is somewhat special. The length of each side is a whole number, and these are the smallest possible whole-number lengths for a right-angled triangle.

Of course, any enlarged version of a 3, 4, 5 triangle will also be right-angled (for example, a 6, 8, 10 triangle).

The next three smallest lengths that make a right-angled triangle are 5, 12 and 13 (check $5^2 + 12^2 = 13^2$). These two types of right-angled triangle are shown in Figure 44.

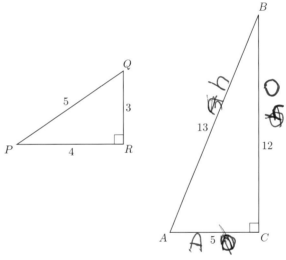

Figure 44   A $(3, 4, 5)$ triangle and a $(5, 12, 13)$ triangle.

Activity 18   *Angles in a* (5,12,13) *triangle*

(a) Use Figure 44 to write down the values of $\cos A$ and $\tan B$ in a $(5, 12, 13)$ triangle. (Leave your answers as fractions.)

(b) Use your calculator to find the sizes of the angles at A and B in radians (correct to 3 decimal places).

3.3 Using trigonometric ratios

In Section 1, you saw how to work out the height of a tall vertical object, such as a tree, by using similar triangles. Here is another way of finding a height, using trigonometry. It has the advantage that you do not have to wait until the Sun comes out! Moreover, it does not require a stick, it can be used where the ground is not flat, and it works for objects, like mountains, whose shadows are not easy to measure.

In order to use this trigonometric method, you have to be able to measure angles. Figure 45 illustrates a simple device for doing this; it is called a *clinometer* (because it measures *inclines*), and is made by modifying a protractor.

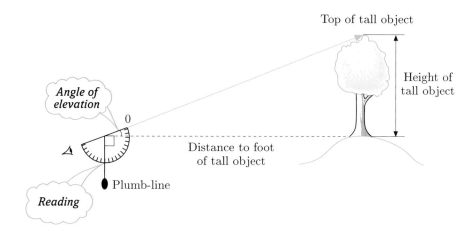

This figure is not drawn to scale.

Figure 45   Using a protractor as a clinometer.

Point the straight side of the protractor towards the tall object and line up the zero with the object's highest point. Make a note of the reading corresponding to the plumb-line. Subtract 90° from this reading, and you will have the angle that the line from your eye to the top of the tall object makes with the horizontal. This angle is called the *angle of elevation* of the object.

To find the height of a tall object by the trigonometric method, you need to know the horizontal distance from the clinometer to the foot of the object, as well as the angle of elevation. Now,

$$\tan\,(\text{angle of elevation}) = \frac{\text{opposite}}{\text{adjacent}} = \frac{\text{height of tall object}}{\text{distance to its foot}}.$$

From this it follows that

height of tall object = distance to its foot × tan (angle of elevation).

You need to be a bit careful though, because the height given by this formula is from the point (approximately eye level) at which the angle of elevation is measured and *not* from ground level.

Also, the foot of the object may not be on the same level as *your* feet.

Example 7   How high is the cliff?

Umberto, who is 2.0 m tall, stands about 70 m from the foot of a vertical cliff and, by using a clinometer, measures the angle of elevation of the top of the cliff as 67° (see Figure 46). How high is the cliff?

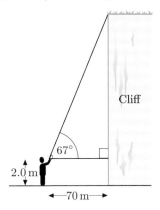

Figure 46   Umberto and the cliff.

Solution

Look at the right-angled triangle in Figure 46. The height, h metres, of the cliff above the top of Umberto's head is given by the side opposite the angle of elevation, while the adjacent side is of length 70 m. Then

$$\tan 67° = \frac{\text{opposite}}{\text{adjacent}} = \frac{h}{70},$$

and so $h = 70 \times \tan 67° \simeq 165$. Therefore, the top of the cliff is approximately 165 m above Umberto's eye level. The height of the cliff is about 2 m greater than this, and so is around 167 m.

However, the accuracy of the angle and distance measurements (which are, in any case, given only to the nearest degree and metre, respectively) will affect the accuracy of this result. The most that can be said is that probably the height of the cliff lies between 160 m and 170 m.

Activity 19   Revisiting the height of a tree

(a) On 21 December in a given year, in Manchester, the angle of elevation of the Sun at its highest point (called its *zenith*) was 13.0°, measured from ground level. At this time, also in Manchester, a particular larch tree gave a shadow 94.0 m long, measured from the centre of its trunk. Assume that the ground was horizontal and the tree rose to a point. How high was the tree?

(b) On 21 March, the following year, in Manchester, the angle of elevation of the Sun at its zenith was 36.5°, again measured from ground level.

At that time, the same tree had a shadow 30.0 m long, measured from the centre of its trunk. How high does this make the tree at that time?

(c) What other assumption has been made? How might it effect the results?

Activity 20   *Handbook activity*

Look back at any notes you jotted down about the trigonometric functions when you were studying *Unit 9* and the associated chapter of the *Calculator Book*, and now amplify these to make a comprehensive Handbook entry on trigonometric ratios.

Example 8   *Map bearings*

You think you are at a point X and you want to walk to a point Y. On your map, Y is 4.9 km north and 3.6 km east of X. What is the map bearing of Y from X?

Solution

The bearing is the angle θ shown in Figure 47. The length of the side opposite θ is 3.6 km, and the length of the adjacent side is 4.9 km. So

$$\tan \theta = \frac{3.6}{4.9}, \qquad \text{and}$$

$$\theta = \tan^{-1}\left(\frac{3.6}{4.9}\right) = 36° \text{ (to the nearest degree)}.$$

Although the bearing is 36°, you would need to correct for magnetic north to get your compass bearing.

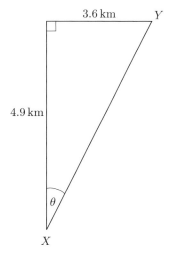

Figure 47

Note that when using trigonometry for bearings, it is sensible to draw a diagram. This is because the angle you have found may not be the actual bearing: you may need to add it to, or subtract it from, 90°, 180° or 270°. Figure 48 shows an example where, to get the bearing, you will have to add 90° to the angle θ.

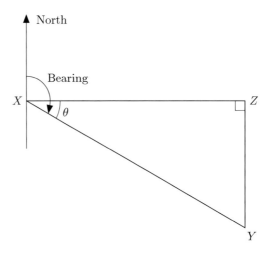

Figure 48

Activity 21   *Getting from A to B*

You are at point A and you want to get to point B. You have located both points on the map. You see that a certain point—point C—is 6.3 km due north of point A and 4.1 km due west of point B. What is the map bearing of B from A? Use your answer to work out how far B is from A. Check your result by using Pythagoras' theorem.

3.4 *Road gradients revisited*

If you think back to *Unit 6* and also to Section 4 of *Unit 9*, you may realize that the ratios used in measuring the steepness of the slope of a road going up a hill are the same as two of the trigonometric ratios. Figure 49 depicts a model of a hill as a right-angled triangle.

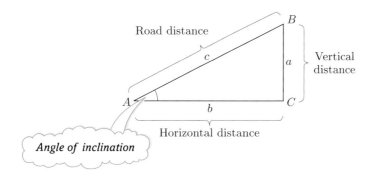

Figure 49   Triangle model of a hill.

It would be reasonable to take the size of the angle A, at which the road is inclined to the horizontal, as the measure of the steepness of the hill. The larger the angle A, the steeper the hill. But it is rather difficult to measure this angle either from a map or from the road itself, and so the two preferred methods of specifying the steepness of a slope avoid the use of this angle. Instead, both methods specify the steepness in terms of the ratio of a pair of sides in the triangle representing the hill. One method uses the ratio of the vertical distance (change in height of the road between two points) to the horizontal distance (measured from a map); this gives the *map gradient*, or *mathematical gradient*. In terms of the symbols in Figure 49,

$$\text{map gradient} = \frac{a}{b} = \frac{\text{opposite}}{\text{adjacent}} = \tan A.$$

The *road gradient*, on the other hand, is the ratio of the vertical distance to the road distance (measured along the surface of the road):

$$\text{road gradient} = \frac{a}{c} = \frac{\text{opposite}}{\text{hypotenuse}} = \sin A.$$

The road gradient may be given as a fraction or a percentage; for example, a road gradient of $1/5$ (1 in 5) is 20%.

The above formulas enable you to convert between the angle of inclination, A, of the hill and either the road gradient or the map gradient, by using trigonometry.

If a hill rises at an angle of $8°$ to the horizontal, then it has a map gradient of $\tan 8° = 0.1405$ (to 4 decimal places); that is, just over 14%. Its road gradient is $\sin 8° = 0.1392$ (to 4 decimal places); that is, just under 14%. For a comparatively small angle of inclination, the difference between the map and road gradients is small, but for larger angles it becomes significant. At an angle of $30°$, the map gradient is about 58% ($\tan 30° \simeq 0.58$), while the road gradient is 50% ($\sin 30° = 0.5$). Few roads are as steep as this!

To find the angle of inclination corresponding to a gradient given as a percentage, it is necessary to use the *inverse function*: $\tan^{-1}$ or $\sin^{-1}$. For example, if the map gradient of a road is given as 1 in 10, or 10%, then the angle of inclination of the road is $\tan^{-1}(0.1) \simeq 5.7°$, and so a road whose map gradient is 1 in 10 makes an angle of just under 6° with the horizontal.

Activity 22   *Road gradients*

(a) Use your calculator to find the gradients (map and road) of a hill that makes an angle of 15° with the horizontal.

(b) Use your calculator to find the angle that a road makes with the horizontal if its map gradient is 15%. What if its *road* gradient is 15%?

(c) Make an entry on map and road gradients on your Handbook activity sheet.

3.5 *Other trigonometric functions*

If you are short of time, skim over this section and Section 3.6.

There are some other trigonometric functions that you may come across. However, you are unlikely to meet them very often.

You may recall from Chapter 9 of the *Calculator Book* that the graph for the cosine function is similar to that for the sine function, but moved across. The reason for this is that the cosine of an angle is actually the sine of a closely related angle. Look at triangle $OP'Q'$ (Figure 50).

You have met this triangle before in the context of an enlargement of the unit circle.

ϕ is the greek letter phi.

Figure 50

Because the angle at Q' is 90°, the angles θ and ϕ must add up to 90° (the sum of the angles in a triangle is 180°). Pairs of angles like these that add up to 90° are called *complementary* angles: ϕ is the complementary angle of θ. Consider the sine of the angle ϕ:

$$\sin \phi = \frac{\text{opposite}}{\text{hypotenuse}} = \frac{x}{r},$$

which is the same as the cosine of the angle θ. The sine of the complementary angle to θ is *cosine* θ or $\cos \theta$.

Hence the name 'cosine'.

In a similar way, the *co*tangent (abbreviated to cot) of the angle θ is defined as the tangent of the complementary angle ϕ, so

$$\cot \theta = \tan \phi = \frac{x}{y}.$$

From this, you can see that

$$\cot \theta = \frac{1}{\tan \theta}.$$

There is another trigonometric ratio that you may meet. This is the secant (abbreviated to sec). In the unit circle, the secant is given by the length, OB, obtained when the radius OP is extended to meet the tangent AB, as shown in Figure 51.

The Latin verb *secare* means 'to cut' (as in 'secateurs' or 'intersect').

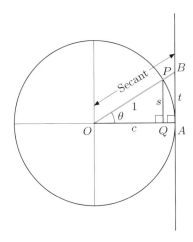

Figure 51 Defining a secant from the unit circle.

Triangle OBA is similar to triangle OPQ in Figure 51. The corresponding lengths in these triangles are set out in the table below.

Triangle OPQ	Triangle OBA
$OP = 1$	$OB = \sec \theta$
$OQ = c$	$OA = 1$

As $OP = 1$, the scale factor that enlarges OP to OB ($= \sec \theta$) must be $\sec \theta$. The same scale factor takes OQ to OA, so

$$OA = OQ \times \sec \theta,$$

or

$$1 = c \times \sec \theta.$$

Hence

$$\sec \theta = \frac{1}{c} = \frac{1}{\cos \theta}.$$

Referring back to the triangle shown in Figure 50, you will recall that $\cos\theta = \dfrac{x}{r}$. Therefore, $\sec\theta$ can be written as

$$\sec\theta = \frac{r}{x}.$$

The *cosecant* (abbreviated to cosec) of the angle θ is the secant of the complementary angle ϕ, and is given by

$$\operatorname{cosec}\theta = \sec\phi = \frac{r}{y}.$$

So

$$\operatorname{cosec}\theta = \frac{1}{\sin\theta}.$$

Summarizing the connections between these trigonometric functions:

$$\cot\theta = \frac{1}{\tan\theta} = \frac{\cos\theta}{\sin\theta}, \quad \sec\theta = \frac{1}{\cos\theta}, \quad \operatorname{cosec}\theta = \frac{1}{\sin\theta}.$$

These functions do not appear on the course calculator. The reason is that, if they are needed, they can be found easily from the other three functions (sin, cos and tan). For example, to find $\operatorname{cosec}40°$, you would find $\sin 40°$ and then press the reciprocal key.

The reciprocal key is labelled x^{-1}.

Inverse and reciprocal

You need to take care not to get confused over two very different uses of the notation $^{-1}$. It is used both for the reciprocal, as in $x^{-1} = 1/x$, and for inverse trigonometric functions, such as $\tan^{-1}$. These have very different meanings.

For reciprocals:

$$2^{-1} = \frac{1}{2}, \qquad \left(\frac{1}{2}\right)^{-1} = 2, \qquad a^{-1} = \frac{1}{a}.$$

However, $\tan^{-1}(10°)$ is **not** $\dfrac{1}{\tan(10°)} = \cot(10°)$. The cotangent is quite a different function from the inverse tangent.

For historical reasons, $\tan^{-1}$ is a common notation for the inverse tangent function. As mentioned earlier, other (less confusing) notations, such as arctan, are sometimes used.

3.6 Trigonometric identities

If you are short of time, skim over this section.

In Figure 50, the complementary angles θ and ϕ add up to $90°$, and so ϕ can be written as $90° - \theta$. The relationships between the trigonometric functions of the complementary angle can then be written as

$$\cos\theta = \sin(90° - \theta), \quad \cot\theta = \tan(90° - \theta), \quad \operatorname{cosec}\theta = \sec(90° - \theta).$$

Such equations are sometimes useful. They are called identities, as they are always true, whatever value the angle θ takes.

Another identity can be obtained from the definition of $\sin\theta$ that is based on the unit circle (see Figure 52). There are pairs of angles that have identical values of $\sin\theta$: for instance, one acute angle θ_1 in the first quadrant and one obtuse angle θ_2 in the second quadrant, as in Figure 52. Because the positions of P_1 and P_2 are symmetrical, $\theta_2 = 180° - \theta_1$, and since $\sin\theta_1 = s = \sin\theta_2$, it follows that $\sin(180° - \theta) = \sin\theta$.

Recall that the sine of an angle is given by the height, s, of P_1 (or P_2) above the horizontal.

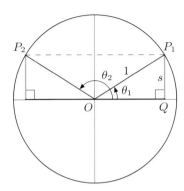

Figure 52   A pair of angles with the same values of $\sin\theta$.

Another useful trigonometric identity comes from expressing Pythagoras' theorem in terms of trigonometry. Think back to the definitions of sine and cosine obtained from the unit circle (see Figure 53 as a reminder).

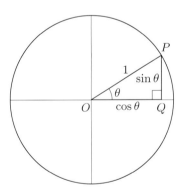

Figure 53   The unit circle showing $\sin\theta$ and $\cos\theta$.

The sides of the right-angled triangle OPQ are $\cos\theta$, $\sin\theta$ and 1. Then, using Pythagoras' theorem,

$$(\cos\theta)^2 + (\sin\theta)^2 = 1.$$

Although this identity has been shown to be true for angles in a right-angled triangle, it is, in fact, true for any size of angle, as are the other identities given above.

$(\cos\theta)^2$ is sometimes written as $\cos^2\theta$, and $(\sin\theta)^2$ as $\sin^2\theta$. So this identity is also written as $\cos^2\theta + \sin^2\theta = 1$.

Activity 23   *Handbook identities*

Add the important trigonometric identities to your Handbook activity sheet, and make a note about the different uses of the $^{-1}$ notation.

Activity 24   *Strategies revisited*

Before leaving Section 3, consider the strategies that you have been using in acquiring skills up to this point in the unit. You might add these strategies to the sheet for Activity 1. In particular, consider how you have been using examples.

Outcomes

After studying this section, you should be able to:

◇   explain the significance of the concept of similarity for the definitions of the trigonometric ratios: sine, cosine and tangent (Activities 17 and 20);

◇   use these ratios to solve problems involving right-angled triangles (Activities 15, 16, 18, 19, 21 and 22);

◇   be aware of trigometric identities (Activity 23);

◇   consider the role of examples in the learning strategies that you have been using (Activity 24).

4 Trigonometry and surveying

Aims   This section aims to extend your repertoire of trigonometric techniques and to demonstrate their use in practical contexts, such as surveying.   ◇

Although maps are often presented on a square grid, until recently surveying was based on a grid of triangles. The Ordnance Survey used triangulation pillars (often called 'trig points') as the vertices of its triangles: one side of a triangle was measured very accurately, and angles at each end of this side were also measured, then trigonometry provided the means of calculating the other two sides. The lengths of these sides could subsequently be used in calculations for neighbouring triangles. This process, also employed in navigation, is called *triangulation*.

Figure 54 shows a typical grid of triangles. The trig points would be at the vertices *A*, *B*, *C*, and so on.

A trig point that is shown on your OS map is on the summit of Win Hill (grid reference 186851). Nowadays, trig points are no longer used; aerial photography and global positioning systems are the main means of surveying.

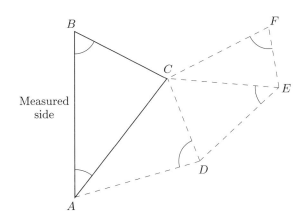

Figure 54   Grid of triangles for surveying.

Once the network of triangles was complete, other objects could be located by measuring their bearings from nearby triangulation points and by again using trigonometry.

In Section 3, you saw how trigonometric methods can be used to find unknown angles and lengths of sides in right-angled triangles. The Ordnance Survey network, on the other hand, contains triangles that are not right-angled. But this does not present a problem because the unknown sides and angles of any triangle can be found by splitting it into two right-angled triangles (as shown in Figure 55 overleaf). However, this is time consuming, and it is more convenient to have general formulas that can be employed in such calculations.

The situation is a little more complicated if the vertices of the triangles are at different heights, but the same principles apply.

There are two commonly used general formulas that apply to any triangle: the sine formula and the cosine formula.

4.1 The sine formula

The *sine formula* relates the lengths of the sides of a triangle to the sines of the angles opposite those sides. The formula is obtained by dividing a general triangle, ABC, into two right-angled triangles, ACD and BCD, as in Figure 55.

When the angle A or B is an obtuse angle (greater than 90°), then the diagram is a little different, but the mathematics is the same. The identity $\sin\theta = \sin(180° - \theta)$ is used.

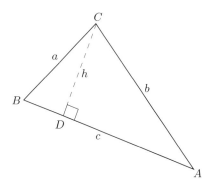

Figure 55

Let CD, the perpendicular height of the triangle, be h. Then in triangle ACD,

$$\sin A = \frac{h}{b}. \tag{2}$$

In triangle CBD,

$$\sin B = \frac{h}{a}. \tag{3}$$

Multiplying both sides of equation (2) by b will give h on its own, as will multiplying both sides of equation (3) by a. So

$$b\sin A = h \quad \text{and} \quad a\sin B = h.$$

When h is eliminated,

$$b\sin A = a\sin B.$$

Dividing both sides of this equation by $\sin A$ and then by $\sin B$ gives

$$\frac{a}{\sin A} = \frac{b}{\sin B}.$$

The triangle in Figure 55 was divided into two right-angled triangles by drawing the perpendicular from C onto AB. However, the perpendicular could have been drawn from any vertex: for example, from A onto BC. In that case, working through the same steps as above, but with A replaced by C, would lead to the formula

$$\frac{b}{\sin B} = \frac{c}{\sin C}.$$

Putting these results together gives the full sine formula:

$$\frac{a}{\sin A} = \frac{b}{\sin B} = \frac{c}{\sin C}.$$

This formula is very useful for finding the length of one side of a triangle when the length of another side and the sizes of two angles are known.

Example 9 *Surveying*

A surveyor chooses two points, A and B, on flat land, such that B is 500 m due north of A. The surveyor then takes the bearings of a number of other locations from A and from B. One of these locations is a church at C. The bearing of C from A is 38°, and that of C from B is 106°. Use the sine formula to calculate the distance of C from A and from B, to the nearest metre.

Solution

The relative positions of A, B and C are shown in Figure 56.

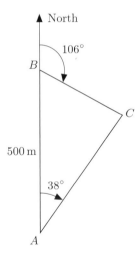

Figure 56

Angle B in triangle ABC is $180° - 106° = 74°$. So angle C is $180° - 74° - 38° = 68°$.

Since the length of $AB (= c\,\text{m}) = 500\,\text{m}$,

$$\frac{c}{\sin C} = \frac{500}{\sin 68°}.$$

Then the sine formula

$$\frac{a}{\sin A} = \frac{b}{\sin B} = \frac{c}{\sin C}$$

can be used to find the other sides.

Recall that side a is opposite angle A, and so on.

Thus

$$\frac{a}{\sin 38°} = \frac{500}{\sin 68°}.$$

This gives

$$a = \frac{500 \sin 38°}{\sin 68°} = 332 \text{ (to the nearest whole number).}$$

Similarly,

$$\frac{b}{\sin 74°} = \frac{500}{\sin 68°}.$$

So

$$b = \frac{500 \sin 74°}{\sin 68°} = 518 \text{ (to the nearest whole number).}$$

Therefore the church at C is 518 m from A and 332 m from B (correct to the nearest metre).

Activity 25   *More surveying*

A surveyor chooses two points, X and Y, on flat land, such that Y is 1 km north of X. The surveyor then takes the bearings of a number of other locations from X and from Y. One of these locations is a windmill at W. The bearing of W from X is 45°, and that of W from Y is 120°. Use the sine formula to calculate the distance of the windmill from X and from Y, to the nearest metre.

A word of caution: although it might seem that the sine formula can be used to find an angle when you know two sides and another angle, it is not usually a suitable method. The reason is that, depending upon the arrangement of lengths and angles, there can be more than one angle, and hence more than one triangle, that will fit the data.

When the given angle is equal to, or greater than, 90°, there is no ambiguity, and you can use the sine formula to find another angle.

It is important to bear in mind that if you know two sides and the *included* angle, you cannot use the sine rule. However, you can use another formula instead: the cosine formula.

4.2   *The cosine formula*

The *cosine formula* is a generalization of Pythagoras' theorem to a triangle that is not right-angled. It is obtained by splitting the triangle into two right-angled triangles, as in Figure 55, and then applying Pythagoras' theorem to each triangle. The derivation of this formula is given in the Appendix and is optional.

There are two forms of the cosine formula (one obtained from the other by algebraic manipulation). For a triangle labelled as in Figure 57, the cosine formula for the side c is given by

$$c^2 = a^2 + b^2 - 2ab\cos C. \tag{4}$$

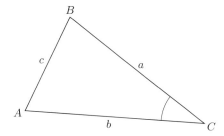

Figure 57

Note that if angle C is a right angle, then $\cos C$ is zero, and the cosine formula reduces to Pythagoras' theorem—as it should!

You can think of the final term $-2ab\cos C$ in equation (4) as a 'correction factor' applied to compensate for the fact that the triangle is not right-angled. In the case of a right-angled triangle, the correction factor is zero.

This is an example of a particular type of test that can often be carried out on a new mathematical result. It confirms that the new result is consistent with what has gone before, and shows that the new result is, indeed, a generalization of the old result; the old result then becomes a special case of the new result.

Example 10   *Find the side*

A triangle has sides of lengths 1 m and 3 m, and the angle between these sides is 30°. Find the length of the side opposite this angle.

Solution

With $a = 1$, $b = 3$ and $C = 30°$, use the cosine formula

$$c^2 = a^2 + b^2 - 2ab\cos C.$$

This gives

$$c^2 = 1 + 9 - 6\cos 30°,$$

so

$$c = \sqrt{10 - 6\cos 30°} = 2.2 \text{ (to 1 decimal place)}.$$

Therefore the length of the side opposite the 30° angle is 2.2 m.

Activity 26 *Find the opposite side*

A triangle has sides of lengths 2 m and 4 m, and the angle between these sides is 60°. Find the length of the side opposite this angle.

The cosine formula, given in equation (4), has the square of one of the sides, c, as the subject. However, if you wish to use the cosine formula to find an angle, it is more convenient to have the cosine of the angle as the subject. This version of the formula can be obtained by rearranging

$$c^2 = a^2 + b^2 - 2ab \cos C$$

as follows.

Add $2ab \cos C$ to both sides, and subtract c^2 from both sides:

$$2ab \cos C = a^2 + b^2 - c^2.$$

Divide both sides by $2ab$ and obtain

$$\cos C = \frac{a^2 + b^2 - c^2}{2ab}. \qquad (5)$$

Although this is a formula for the angle C, the cosine formula can be applied to *any* of the angles of a triangle when the lengths of all the sides are known.

The version of the cosine formula in equation (5) can be written in words in such a way as to make it clear how to apply the formula to any angle in a triangle (see Figure 58):

> The cosine of the angle required is the sum of the squares of the lengths of the two adjacent sides *minus* the square of the length of the opposite side, all divided by twice the product of the lengths of the two adjacent sides.

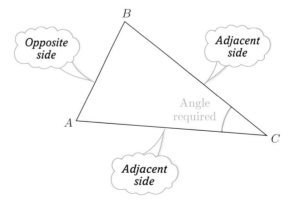

Figure 58

Example 11   *Formula for* $\cos A$

Write down the cosine formula with $\cos A$ as the subject, by using the verbal formula given above Figure 58 on the opposite page.

Solution

If the required angle is A, then the adjacent sides are b and c and the opposite side is a. Using this information in conjunction with the verbal formula gives

$$\cos A = \frac{c^2 + b^2 - a^2}{2cb}.$$

This result can also be obtained by changing C to A in equation (5), and then replacing every c by a and every a by c in that equation.

Activity 27   *From other points of view*

Now use a similar method to obtain the formula for $\cos B$.

Example 12   *Find the angle*

A triangle has sides of lengths $2\,\mathrm{m}$, $3\,\mathrm{m}$ and $4\,\mathrm{m}$. Find the angle opposite the shortest side.

Solution

With the distances in metres, let $a = 2$, $b = 3$ and $c = 4$. Using the formula

$$\cos A = \frac{c^2 + b^2 - a^2}{2cb} = \frac{4^2 + 3^2 - 2^2}{24}$$

gives

$$A = \cos^{-1}\left(\frac{4^2 + 3^2 - 2^2}{24}\right) = 29° \text{ (to the nearest degree)}.$$

The cosine formula is rather tedious to use if you have to keep applying it over and over again. But it can easily be programmed for the calculator. To save time, you can use the cosine formula program, details of which you will find in the *Calculator Book*.

Work through Section 14.1 of the Calculator Book.

Below is an example of a practical situation in which the program or the formula itself can be used to find angles.

Example 13   *Garden planning with trigonometric tools*

Suppose that you are thinking of making some improvements to your garden. You need an accurate scale plan of the garden, showing the positions of various items that you do not want to move or cannot move. You do not have any surveying equipment, so measuring angles in the garden will not be very easy—your protractor is rather small! Measuring distances with the aid of your tape measure, on the other hand, is quite straightforward. Therefore you can measure the distances between the relevant objects in the garden. However, when it comes to drawing the plan, it would be helpful to know the angles.

Figure 59 shows a sketch of the various distances you have measured—but this is by no means a reliable scale plan of the garden. Use trigonometry to find angle A in triangle ABD.

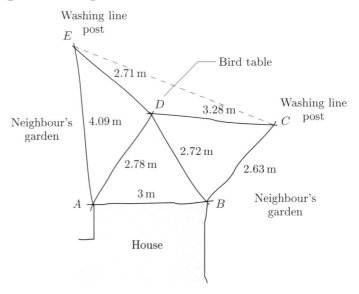

Figure 59   Rough sketch plan of the garden.

Solution

To find angle A in triangle ABD, you can use the cosine formula program or the formula itself directly:

$$\cos A = \frac{2.78^2 + 3^2 - 2.72^2}{2 \times 2.78 \times 3}.$$

So

$$\text{angle } A = \cos^{-1}\left(\frac{2.78^2 + 3^2 - 2.72^2}{2 \times 2.78 \times 3}\right) = 55.988\,968\,13° \simeq 56°.$$

It is pointless to give the answer to greater accuracy than you could hope to measure with a protractor when drawing your scale plan.

Activity 28   *Using the cosine formula*

(a) Find the angles ABD and ADB in Figure 59.

(b) You are now in a position to locate point D exactly on the plan of the garden, and to draw in the precise positions of the lines AD and BD (assuming that you have already drawn AB, the side of the house, which serves as a baseline). To complete your plan you need to find some of the angles of triangles ADE and BCD. Find these angles, and hence complete the plan of the garden.

4.3   Letter from Ireland

The audio band for this unit contains an extract from a radio broadcast. It describes the use of trigonometric, geometric and other mathematical concepts in a variety of contexts. As you listen, sketch diagrams of the situations being described and think about how the concepts of similarity and trigonometry might be used in the calculations discussed.

Now listen to band 2 of CDA5510 (Track 7).

Activity 29   *Explaining an Irish idea*

The speaker on the audio, Tim Robinson, says that he could find the height of either Cashel Hill or Derryclare Mountain by knowing the angles of elevation of each hill from two points on the road, 1000 paces apart. Figure 60 summarizes this information for Cashel Hill (assuming that 1 pace is 1 m). By considering each of the two right-angled triangles in the diagram, explain how to find the height of Cashel Hill.

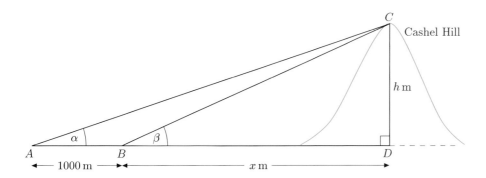

Figure 60   Model of Cashel Hill.

This section has shown how to find unknown lengths and angles in triangles that do not have a right angle.

There are six key measurements in a triangle: three sides and three angles. If you know three or more of these measurements, then the sine and cosine formulas can often help you to find the others. When you know

- two angles and one side, use the sine formula;
- all three sides, use the cosine formula;
- two sides and the included angle, use the cosine formula.

When you know two sides and an acute angle opposite one of them, you cannot reliably find the other angles.

If you only know angles and not sides, these formulas will not help as only the shape of the triangle is specified and not the size.

Outcomes

After studying this section, you should be able to:

◇ use the cosine and sine formulas to solve problems involving triangles, including some related to surveying (Activities 25, 26, 28 and 29);

◇ rearrange the cosine formula, and rewrite it for differently named angles and sides (Activities 27 and 28).

5 Showing the way they went

Aims   This section aims to consolidate and build upon some of the topics from earlier in the course. It brings together ideas about maps, similarity and trigonometry, and uses them in the context of local and global journeys.   ◇

5.1 Maps and journeys

Activity 30   *Maps and grid references*

The video band you are about to watch begins by revisiting the walk from the *Unit 6* video band. Look at the OS map and find Mam Tor (grid reference 127836) at the beginning of the ridge along which the walkers went. From there they walked along the ridge to Hollins Cross. Write down the grid reference of Hollins Cross, and find the distance on the map between it and Mam Tor.

If necessary, look back at your notes for *Unit 6* on grid references, on converting map distances into ground distances (and vice versa), and on calculating distances between points from their grid references.

Now watch band 11a on DVD00107, 'Showing the way they went'. As you watch, think about how this balloon journey would be represented on a map.

For planning and navigating balloon flights, it is crucial to know the wind velocity. The commentary on the video suggested that the wind forecast was not accurate in every way. To check whether the actual balloon flight was consistent with the forecast, you have to calculate the average velocity for the flight. In order to do this you need to know the distance and bearing of the balloon journey, as well as the flight time. You could measure the bearing with a protractor directly from the map, and you could obtain the distance travelled by measuring the distance on the map and multiplying by the map scale. However, you do not need to use the map, as you can calculate both the distance and the bearing from the grid references. The distance can be obtained by using Pythagoras' theorem, and the bearing can be obtained by using trigonometry. Example 14 overleaf shows how this is done.

Remember that velocity (discussed in *Unit 11*) includes both the magnitude (speed) and the direction (bearing).

Example 14   *Reconstructing a balloon journey*

The first balloon took off from grid reference 122825 and landed at 150888, 45 minutes later.

(a) Use the grid references and Pythagoras' theorem to find the total distance travelled.

(b) Use trigonometry to find the bearing of the landing site from the take-off site. Assume that the correction for magnetic north was $5°$ in 1995, when the flight took place.

(c) Hence find the average velocity (speed and direction) of the balloon.

(d) The forecast wind velocities were $8\,\text{km}$ per hour at a bearing of $25°$ up to 2000 feet, increasing towards $14\,\text{km}$ per hour at a bearing of $90°$ at higher altitudes. The balloon flew at altitudes from about 400 to 3000 feet. Was the wind forecast consistent with the actual flight?

Solution

(a) As explained in Section 6.2 of the *Calculator Book*, Pythagoras' theorem can be used to find distances from grid references by separating the grid references into eastings and northings, since these represent the x- and y-coordinates in the plane. The squares of differences between the eastings and northings of the landing and take-off sites can be found, as in Table 1.

Table 1

	Grid reference	Easting	Northing
Take-off	122825	122	825
Landing	150888	150	888
Differences		$150 - 122 = 028$	$888 - 825 = 063$
Square of differences		784	3969

Then Pythagoras' theorem gives the map distance as $\sqrt{784 + 3969}$ units.

Each single grid reference unit represents $0.1\,\text{km}$ on the ground. So the distance is

$$\sqrt{784 + 3969} \times 0.1\,\text{km} \simeq 68.9 \times 0.1\,\text{km} = 6.9\,\text{km}$$
$$\text{(to 2 significant figures).}$$

(b) In this case, the tangent of the grid bearing is given by the difference between the eastings divided by the difference between the northings (see Figure 61). From the table above, this is $28/63$. So the grid bearing is $\tan^{-1}(28/63) = 24°$ (correct to the nearest degree). Adding the correction for magnetic north gives $(24 + 5)° = 29°$.

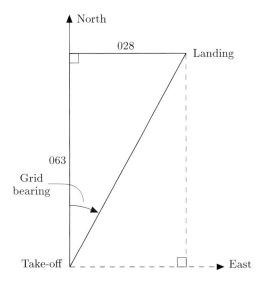

Figure 61

(c) The journey took 45 minutes, which gives an average speed of

$$6.9 \, \text{km}/0.75 \, \text{hour} = 9.2 \, \text{km per hour}.$$

Therefore the average velocity of the balloon during the flight was 9.2 km per hour at a bearing of 29°.

(d) As the balloon was flying both above and below 2000 feet, from the forecast you would expect its average velocity to be somewhere between 8 and 14 km per hour and at a bearing between 25° and 90°. The actual balloon velocity calculated in part (c) fits this. So the balloon flight was consistent with the wind forecast.

Activity 31 *Stressing and ignoring*

Remember from the video that Chris, the pilot of the second balloon, recorded his map positions from his GPS (global positioning system) device and used this information to assess his fuel situation before passing the point of no return over the moors. Chris noted his map position at different times along his journey, as shown in Table 2. Thus, his journey can be represented by a series of grid references. What does this representation stress and what does it ignore?

Table 2

Time	Grid reference	Notes
9.40	123825	Take-off
10.26	157885	See other balloon land
10.53	184927	Last pre-moors landing opportunity
11.02	193940	Moors
11.14	209957	Moors
11.40	232998	Landing

Activity 32   *Reconstructing Chris's balloon journey*

Chris's grid references and times, given in Table 2 on the preceding page, can be used to calculate the distances and speeds for the various legs of his trip. This is done for the first leg of his trip in Table 3. Complete Table 3 by calculating the distances and speeds for the second and third legs.

Table 3

Leg	Grid ref. 1	Grid ref. 2	Distance east	Distance north	Distance as crow flies/km	Time taken/h	Speed/km/h (to 2 s.f.)
1	123825	157885	$157 - 123$ $= 034$	$885 - 825$ $= 060$	$0.1\sqrt{1156 + 3600}$ $= 6.896$	$46/60$ $= 0.767$	9.0
2	157885	184927					
3	184927	193940					

Activity 33   *Reconstructing more of Chris's balloon trip*

Use Chris's grid references from Table 2, in conjunction with trigonometry, to calculate the grid bearings for the second and third legs of his trip, filling in Table 4 as you go. The first leg of the trip has already been completed in the table. Hence find the magnetic bearings for the second and third legs of the journey.

Table 4

Grid ref. 1	Grid ref. 2	Distance east	Distance north	$\tan\theta$	Grid bearing (to nearest degree)	Magnetic bearing (to nearest degree)
123825	157885	$157 - 123$ $= 034$	$885 - 825$ $= 060$	$34/60$	$30°$	$35°$
157885	184927					
184927	193940					

In Activities 32 and 33, both of the distances (east and north) were positive and so the bearing was in the first quadrant. This is not always the case. When one or both of x and y are negative, then the bearing will be in a different quadrant (west and/or south, instead of east and north). Moreover, because the tangent function has two cycles in every full turn, even when you know the value of $\tan\theta$, there are two possible values of θ. To avoid mistakes, it is important to draw a diagram to ascertain the relationship between the angle given by your calculator and the actual bearing.

5.2 *Global journeys*

The next part of the video looks at global journeys, where the distance between two points needs to be calculated as the arc of a circle rather than as a straight line. The shortest distance between two points on the globe is the *great circle* distance, which is along the arc of a circle whose radius is that of the globe. Trigonometry is vital for calculating great circle distances, and the video shows how this is done.

As you watch the video, think about how distances between points are calculated and try to follow the procedure used in finding the lengths of the sides of the triangles. Do not worry if you do not follow all of the details.

Now watch band 11b on DVD00107.

The general formula for calculating distances along great circle routes between points on the Earth whose latitudes and longitudes are known is derived in an optional Appendix. It can be programmed into the calculator. The *Calculator Book* gives the formula and program details.

Now work through Section 14.2 of the Calculator Book.

Recall from Section 1 that the length of the arc of a circle, radius r, that subtends an angle θ, is $l = r\theta$.

If you are short of time, watch the video but omit Activity 34.

$Deg + Time/60.$

Activity 34 *How far to avoid the restricted zone? (Optional)*

In the video, you saw that the Breitling Orbiter balloon was not able to travel along the great circle over the south of China, as it had to avoid flying into the restricted zone north of 26°N latitude. Instead it flew along the 25.5 N latitude, from longitude 98.1°E to longitude 119.9°E.

(a) What is the angle between these two longitudes in radians?

(b) Take the radius, R, of the Earth as 6371 km. As Figure 62 indicates, the radius of latitude 25.5 N will be

$$r = R\cos 25.5°.$$

Calculate r.

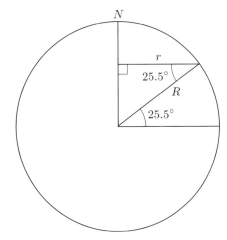

Figure 62

(c) Use the formula $l = r\theta$ to find the distance actually flown by the balloon, to the nearest km.

(d) Use the calculator program for great circle distances to find the great circle distance between the start and finish of this leg of the flight. How does this compare with the distance actually flown?

Outcomes

After studying this section, you should be able to:

◇   use Pythagoras' theorem and trigonometry to calculate distances and bearings of journeys, directly from grid references for the route (Activities 30, 31, 32 and 33);

◇   use the formula $l = r\theta$ to calculate distances on global journeys, with the help of a calculator program (Activity 34).

Unit summary and outcomes

This unit has discussed aspects of art, and has also looked at the physical world of maps and navigation, exploring the calculation of distances and bearings. It has considered ways of seeing and representing in both maps and paintings. One linking notion throughout has been the mathematical concept of similarity. The named trigonometric ratios (sine, cosine, tangent, and so on) rely on the similarity properties of right-angled triangles for their specification. The theory behind perspective painting is also based upon the ideas of similarity. These ideas are very powerful: they underlie techniques for calculating global distances, and also for reconstructing a room depicted in a picture painted over 300 years ago.

Activity 35 *A skills audit*

There is an activity sheet for this audit. Before you finish with the unit, make an assessment of how you feel your skills have improved since you started the course. *Units 5* and *9* included skills audits. Look back at these audits. Were you able to improve and develop the aspects you had identified as needing work? How do you know? What could you use as evidence of your improvement? Are there other skills you have developed and improved?

Outcomes

Now you have finished this unit, you should be able to:

◇ decide whether two figures (in particular, triangles) are similar;

◇ use the fact that two figures are similar to deduce other information about their sides and/or angles;

◇ use the formula $l = r\theta$ for the lengths of arcs of circles;

◇ prove Pythagoras' theorem by using similar triangles;

◇ understand how similar triangles apply to perspective drawing;

◇ explain the significance of the concept of similarity for the definitions of the trigonometric ratios, and use them to solve problems involving right-angled triangles;

◇ use the cosine and sine formulas to solve problems that involve triangles, including some relating to surveying;

◇ use Pythagoras' theorem and trigonometry to calculate distances and bearings of journeys directly from grid references;

◇ identify the strategies and mathematical skills needed for particular topics and recognise mathematical techniques in written accounts.

Appendix: Cosine formulas

The material in this Appendix is *optional*.

1 The cosine formula for a flat triangle (in a plane)

Figure 63 shows a triangle ABC. The cosine formula relates the lengths of the sides a, b and c to one of the angles of the triangle, say, C.

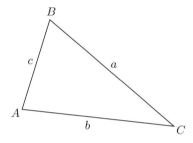
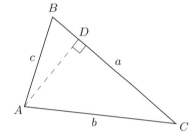

Figure 63

To find the cosine formula, triangle ABC is split into two right-angled triangles, as shown in Figure 63: the two triangles are formed by drawing a perpendicular line from vertex A to a point D on the opposite side. Now AD is a side that is common to both of the right-angled triangles, ADC and ABD. By using Pythagoras' theorem in each of these triangles in turn, and by taking advantage of the fact that AD is common to both triangles, an equation involving a, b, c and $\cos C$ can be produced. Let the length of AD be x.

In triangle ADC, the length of CD is $b \cos C$. Then, using Pythagoras' theorem,

$$x^2 + (b \cos C)^2 = b^2,$$

and so

$$x^2 = b^2 - (b \cos C)^2.$$

In triangle ABD, the length $BD = BC - CD = a - b \cos C$. Therefore, Pythagoras' theorem gives

$$x^2 + (a - b \cos C)^2 = c^2,$$

and so

$$x^2 = c^2 - (a - b \cos C)^2.$$

This yields two expressions equal to x^2. Equating these expressions gives

$$b^2 - (b \cos C)^2 = c^2 - (a - b \cos C)^2.$$

Expanding the bracket on the right-hand side then gives

$$b^2 - (b \cos C)^2 = c^2 - [a^2 - 2ab \cos C + (b \cos C)^2]$$
$$b^2 - (b \cos C)^2 = c^2 - a^2 + 2ab \cos C - (b \cos C)^2.$$

Adding $(b \cos C)^2$ to both sides gives

$$b^2 = c^2 - a^2 + 2ab \cos C.$$

Adding a^2 to both sides and subtracting $2ab \cos C$ from both sides gives

$$a^2 + b^2 - 2ab \cos C = c^2.$$

So

$$c^2 = a^2 + b^2 - 2ab \cos C.$$

2   The great circle formula

The video showed how to calculate the great circle distance between two particular points, S and W, on the surface of the globe. A general formula can be obtained by the same method, replacing the numerical values of the latitudes and longitudes of the points by symbols: the latitudes of S and W by $LT1$ and $LT2$, and their longitudes by $LG1$ and $LG2$, respectively.

In Figure 64, the point O is the centre of the globe, and N is the North Pole. The lines of longitude for S and W are drawn. Tangents are drawn, at the North Pole, to both lines of longitude. The radii OS and OW are extended to intersect with these tangents at S' and W', respectively.

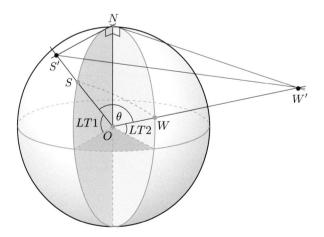

Figure 64

This Appendix does not give the full details of the derivation of the great circle formula, but gives the various steps needed and the intermediate results. The steps needed are:

To find the arc length of the great circle. This arc length can be found using the formula $l = r\theta$ once the angle θ is known ($r = R$, the radius of the Earth).

To find the angle θ. This angle is in the triangle $OS'W'$. It can be found using the cosine formula in this triangle once the sides of the triangle are known. These sides are OS', OW' and $S'W'$.

To find the lengths of the sides OS', OW' and $S'W'$ of triangle $OS'W'$.

(a) Each of the sides OS' and OW' is also in another triangle: OS' is in triangle $OS'N$, which is right-angled; and OW' is in triangle $OW'N$, which is also right-angled. Hence the trigonometric ratios can be used to find the lengths, by referring to Figure 65.

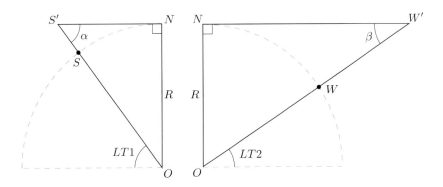

Figure 65

(b) The third length, $S'W'$, is also in triangle $NS'W'$.

This triangle is not right-angled and to find the length $S'W'$ the cosine formula will be needed. This will require finding two sides, NS' and NW', and using their included angle γ in triangle $NS'W'$ (see Figure 66). The angle, γ, can be found from the difference in the longitudes of position S and position W.

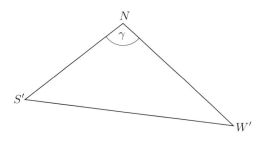

Figure 66

To find the lengths of the sides NS' and NW' of triangle $NS'W'$.
The sides, NS' and NW', can be found from the right-angled triangles in Figure 65 (see previous page).

These steps have to be carried out in the reverse order as the results from each step in the list above are needed by the previous step.

Thus the results are:

Finding the lengths of the sides NS' and NW' of triangle $NS'W'$.

From triangle $OS'N$,
$$NS' = \frac{R}{\tan \alpha} = \frac{R \cos \alpha}{\sin \alpha}.$$
From triangle $OW'N$,
$$NW' = \frac{R}{\tan \beta} = \frac{R \cos \beta}{\sin \beta}.$$

Finding the lengths of the sides OS', OW' and $S'W'$ of triangle $OS'W'$.

(b) From triangle $NS'W'$,
$$(S'W')^2 = \left(\frac{R \cos \alpha}{\sin \alpha}\right)^2 + \left(\frac{R \cos \beta}{\sin \beta}\right)^2 - 2R^2 \frac{\cos \alpha \cos \beta}{\sin \alpha \sin \beta} \cos \gamma.$$

(a) From triangle $OS'N$,
$$OS' = \frac{R}{\sin \alpha}.$$
From triangle $OW'N$,
$$OW' = \frac{R}{\sin \beta}.$$

Finding the angle θ. This is obtained by using the cosine formula in triangle $OS'W'$:
$$\theta = \cos^{-1}(\sin \alpha \sin \beta + \cos \alpha \cos \beta \cos \gamma).$$

However, from Figure 65, $\alpha = LT1$ and $\beta = LT2$, and, as γ is given by the difference in longitudes, $\gamma = LG2 - LG1$. So

θ needs to be in radians at this point.

$$\theta = \cos^{-1}[\sin LT1 \sin LT2 + \cos LT1 \cos LT2 \cos(LG2 - LG1)].$$

Finding the arc length of the great circle. This is given by
$$l = R\theta = R \cos^{-1}[\sin LT1 \sin LT2 + \cos LT1 \cos LT2 \cos(LG2 - LG1)].$$

Note
Sometimes the formula for the arc length is derived in a slightly different form, using
$$\theta = 90° - \sin^{-1}[\sin LT1 \sin LT2 + \cos LT1 \cos LT2 \cos(LG2 - LG1)],$$

rather than the formula for θ given above. But the two formulas for θ are equivalent as $\cos \theta = \sin(90° - \theta)$. You may have noticed this alternative expression for θ was used in Don Campbell's program in the video.

Comments on Activities

Activity 1

You will be building upon geometric ideas and techniques from the preparatory material, as well as upon ideas of similarity from *Unit 2*, scale and map reading from *Unit 6*, trigonometry from *Unit 9*, and proportion from *Unit 13*. You will also be building upon your calculator skills.

It will be quicker if you merely review your notes on these topics, but if you have difficulty with a topic you may also want to look it up in the relevant unit. During your study of this unit you may want to make notes on some of the following: the use of a particular strategy to solve a mathematical problem; the ways in which you present your work; methods of reading text for learning. Strategies relating to how you learn are perhaps more difficult to define clearly. They may include using the modelling cycle; stressing and ignoring; identifying what you already know and what you want to find out; looking at generalities and at specific cases; annotating or highlighting the unit text as you study; and making notes on important terms and techniques.

Activity 2

(a) Not similar: the scale factor for the shorter sides is 2, but that for the longer sides is 1.5.

(b) Similar: all circles are the same shape.

(c) Similar: the information given is enough to identify the shapes as squares, and all squares are similar.

(d) Similar: the angles are all the same and the smaller rectangle can be enlarged by a scale factor of 3 to give the larger rectangle.

Activity 3

Since the scale factor is 2.8, the length $y = 3 \times 2.8 = 8.4$.

Activity 4

(a) Similar: two pairs of corresponding angles of the triangles are the same.

(b) Not necessarily similar: although corresponding sides are in the same ratio and two corresponding angles are equal, those corresponding angles are not the included angles.

(c) Not similar: the three sides are not in the same ratios in both triangles.

(d) Similar: a pair of sides is in the same ratio in both triangles, and the included angles are the same.

Activity 5

The tree has been modelled by what is, in effect, a straight line, and the Sun's rays are represented by parallel straight lines (does light travel in straight lines?). Both the tree and the stick are (explicitly) assumed to be vertical and the ground to be horizontal. It would not matter if the stick and the tree were not vertical or the ground were not horizontal. Provided the angle between the stick and the ground is the same as that between the tree and the ground, the triangles are still similar, as depicted in Figure 67.

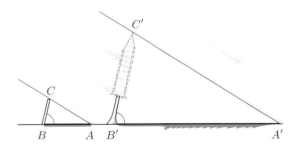

Figure 67

It is also assumed that the stick and the shadows can be measured exactly. Any errors will produce scaled-up errors in the tree measurement. The tree is assumed to have a single height, and the top of the tree's shadow is deemed to correspond to the top of the tree, as shown in Figure 68(a). A bushy tree, as in Figure 68(b), can result in an underestimation of the height of the tree.

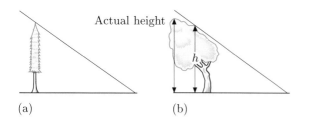

(a)   (b)

Figure 68

Activity 6

Using the annotation from Figure 15,

$$a' = \frac{ac'}{c} = \frac{1.2 \times 6.4}{0.8}\,\text{m} = 9.6\,\text{m}.$$

Activity 7

Similar triangles are created, and ratios (or scale factors) are used.

Activity 8

(a) Two angles in triangle ACD are equal to two angles in triangle ABC: one right angle (at D and C, respectively) and one angle α (at A). This corresponds to the second criterion for similarity in Section 1.2.

(b) In triangle ADC, the two sides with ratio x/b are the hypotenuse and the shortest side. The corresponding sides in triangle ABC are of lengths c and b, respectively, so the corresponding ratio in triangle ABC is b/c.

Activity 9

The ratio of the corresponding sides in triangle ABC is a/c.

As the triangles ABC and CBD are similar, the ratios of the corresponding sides are the same. So

$$\frac{a}{c} = \frac{y}{a},$$

and hence, multiplying through by a gives

$$y = \frac{a^2}{c}.$$

Now substitute for x and y in the equation

$$c = x + y,$$

and obtain

$$c = \frac{b^2}{c} + \frac{a^2}{c}.$$

Multiplying both sides by c gives

$$c^2 = a^2 + b^2.$$

Activity 10

The radius of the Earth, according to Eratosthenes' data, is $20\,000/\pi \simeq 6000\,\text{km}$ (to 1 s.f.).

There are assumptions about the Sun's rays being straight lines, about them being parallel to one another when they arrive at the Earth, and about the Earth being spherical.

Activity 11

Your notes should include the concepts of the picture plane, viewpoint, vanishing point, horizon and distance points.

Activity 12

Your answer will depend upon your environment and your viewpoint. However, the shapes that you see should change when your viewpoint changes. Vertical lines generally remain vertical, but horizontal lines change; for example, the tops and bottoms of doors and windows will only appear parallel when viewed face on, not at an angle.

Activity 13

The process of reconstructing the room is the reverse of constructing the perspective painting. The distance of the artist from the picture plane can be found using the distance points from the painting. A scale factor for an object in the picture also needs to be found; there is always some ambiguity about this but the height of a person could be used to find it. From this information, a plan of the actual room can be produced.

Activity 14

Your notes may build upon those for Activity 11 and may involve diagrams like Figures 26, 28 and 29.

The reason why drawn lines of equal length do not always represent real lines of equal length is that lines different distances from the viewpoint will have different scale factors (the scale factor is given by $k = d'/d$ and will vary depending on the distance d).

Activity 15

(a) Let the opposite side be measured in cm. Then

$$\sin \pi/6 = \frac{\text{opposite}}{\text{hypotenuse}} = \frac{\text{opposite}}{10}.$$

So

$$\text{opposite} = 10 \sin \pi/6 = 5.$$

Length of opposite side is 5 cm.

(b) Let the hypotenuse be measured in cm. Then

$$\sin \pi/4 = \frac{8}{\text{hypotenuse}}.$$

So

$$\begin{aligned}
\text{hypotenuse} &= \frac{8}{\sin \pi/4} \\
&= 11.3137085 \\
&= 11.3 \text{ (to 3 s.f.)}.
\end{aligned}$$

Length of hypotenuse is 11.3 cm.

Activity 16

(a) $\sin \theta = \dfrac{\text{opposite}}{\text{hypotenuse}} = \dfrac{6}{10} = 0.6.$

Then

$$\begin{aligned}
\theta &= \sin^{-1}(0.6) \\
&= 36.86985765° \\
&= 37° \text{ (to the nearest degree)}.
\end{aligned}$$

(b) $\sin \theta = \dfrac{5}{8} = 0.625.$

Then

$$\begin{aligned}
\theta &= \sin^{-1}(0.625) \\
&= 0.6751315329 \\
&= 0.68 \text{ (to 2 s.f.)}.
\end{aligned}$$

Activity 17

The triangles OPQ and $OP'Q'$ are similar. The corresponding lengths in these triangles are summarized in the table below.

Triangle OPQ	Triangle $OP'Q'$
$OP = 1$	$OP' = r$
$OQ = c$	$OQ' = x$

The scale factor is r, thus

$$x = r \times c = r \cos \theta.$$

Therefore

$$\cos \theta = \frac{x}{r}.$$

Activity 18

(a) In triangle ABC,

$$\cos A = \frac{5}{13}$$

and

$$\tan B = \frac{5}{12}.$$

(b) $A = \cos^{-1}(5/13) = 1.176005707$

$$= 1.176 \text{ (to 3 d.p.)}$$

and

$$B = \tan^{-1}(5/12) = 0.394791197$$

$$= 0.395 \text{ (to 3 d.p.).}$$

As a check, if you add angles A and B, you should get $\pi/2$ ($\simeq 1.570796327$).

Activity 19

(a) In the absence of any other information, assume that the tree was vertical. This assumption leads to the diagram in Figure 69, where h m is the height of the tree (the other labels on the diagram are for future reference).

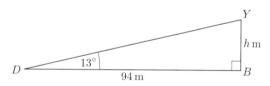

Figure 69   December tree.

Then the tree's height, h m, is given by

$$h = 94 \tan 13° = 21.70160997.$$

The tree was therefore approximately 21.7 m high.

(b) In this case, the information leads to the diagram in Figure 70.

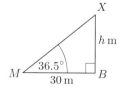

Figure 70   March tree.

This time,

$$h = 30 \tan 36.5° = 22.19883225.$$

According to this calculation, the tree was about 22.2 m high.

(c) It was stated that the ground was horizontal, but it has been assumed without justification that the tree was thin and vertical.

Now, you may have noticed that the tree appears to have grown by about half a metre between December and March! Growth of this amount in three comparatively cold months seems improbable. A possible explanation is that the tree was not vertical. If this was true, the information in parts (a) and (b) can be combined. Assuming that the direction of the shadow cast on both dates was the same, you can draw a single diagram. As the apparent height of the tree on the March date was higher than in December, the diagram must look like Figure 71.

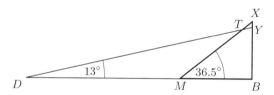

Figure 71

The intersection point of the two lines DY and MX must be the top of the tree, T. This is to the left of the vertical from the base of the tree, so the tree must be leaning towards the shadows. The height of T is less than the height of X or Y. Therefore, the height of the tree is likely to have been overestimated: it may only be about 21 m high.

Activity 20

You might include a diagram like Figure 39, and the definitions

$$\sin \theta = \frac{y}{r} = \frac{\text{opposite}}{\text{hypotenuse}},$$

$$\cos \theta = \frac{x}{r} = \frac{\text{adjacent}}{\text{hypotenuse}},$$

$$\tan \theta = \frac{\sin \theta}{\cos \theta} = \frac{y}{x} = \frac{\text{opposite}}{\text{adjacent}}.$$

Activity 21

The points are shown as the vertices of a right-angled triangle in Figure 72.

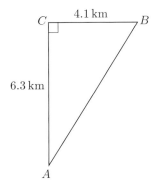

Figure 72

The size of the angle at A in the triangle is

$$A = \tan^{-1} \left(\frac{4.1}{6.3} \right) = 33.055\,822\,81°.$$

So the map bearing of B from A is roughly 33°.

The distance of B from A can be found by calculating either $4.1/\sin A$ or $6.3/\cos A$, and comes to 7.5 km (to 2 s.f.). According to Pythagoras' theorem, the distance (in km) is

$$\sqrt{4.1^2 + 6.3^2} \simeq 7.5.$$

Activity 22

(a) The map gradient is $\tan 15° \simeq 0.27$; that is, about 27%. The road gradient is $\sin 15° \simeq 0.26$; that is, about 26%.

(b) If the map gradient is 15%, then the angle that the road makes with the horizontal is

$\tan^{-1} 0.15 \simeq 8.5°$. If the road gradient is 15%, then the angle is $\sin^{-1} 0.15 \simeq 8.6°$.

(c) In your entry, mention the relationship between the map gradient and $\tan A$; and between the road gradient and $\sin A$.

Activity 23

You should add to your Handbook activity sheet such identities as

$$\cos \theta = \sin(90 - \theta),$$

$$\sec \theta = \frac{1}{\cos \theta}$$

and

$$(\cos \theta)^2 + (\sin \theta)^2 = 1,$$

and, if you have not already got it,

$$\tan \theta = \frac{\sin \theta}{\cos \theta}.$$

Activity 24

Often examples precede an activity that is based on a similar concept. Examples can also play a key role in demonstrating connections, for instance in showing the connection between techniques and definitions. Thus, Example 7 showed the link between the definition of the tangent function and the technique of finding a height from an angle of elevation and a horizontal distance.

Examples can also serve to introduce an idea or to illustrate the meaning of a result. So whenever you meet an example, it is worth considering how you can make the most use of it.

Activity 25

The relative positions of X, Y and W and the angles of the triangle XYW are sketched in Figure 73. The angle at Y is $180° - 120° = 60°$; the angle at W is $180° - (45° + 60°) = 75°$.

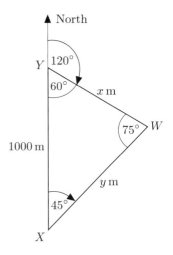

Figure 73

Since the length of XY is $1\,\text{km}\,(= 1000\,\text{m})$, using the sine formula gives

$$\frac{x}{\sin 45°} = \frac{y}{\sin 60°} = \frac{1000}{\sin 75°}.$$

Therefore,

$$x = 1000 \times \frac{\sin 45°}{\sin 75°}$$
$$= 732 \text{ (to the nearest whole number)}$$

and

$$y = 1000 \times \frac{\sin 60°}{\sin 75°}$$
$$= 897 \text{ (to the nearest whole number)}.$$

So the windmill is $732\,\text{m}$ from Y and $897\,\text{m}$ from X (to the nearest metre).

Activity 26

Make $a = 2$, $b = 4$ and the angle $C = 60°$.

Then using the cosine formula,

$$c^2 = a^2 + b^2 - 2ab \cos C$$
$$= 4 + 16 - 2 \times 2 \times 4 \times \cos 60°$$
$$= 4 + 16 - 2 \times 2 \times 4 \times 0.5$$
$$= 12.$$

So $\quad c = \sqrt{12} = 3.5$ (to 1 d.p.).

The length of the third side is $3.5\,\text{m}$ (to 1 d.p.).

Activity 27

Interchanging B and C, and also b and c gives

$$\cos B = \frac{a^2 + c^2 - b^2}{2ac}.$$

Activity 28

(a) To find the angle at B in triangle ABD, use the cosine formula:

$$\cos B = \frac{3^2 + 2.72^2 - 2.78^2}{2 \times 3 \times 2.72}$$
$$= \frac{8.67}{16.32} = 0.531\,25.$$

So angle B is

$$\cos^{-1} 0.531\,25 = 57.910\,048\,74 \simeq 58°.$$

(Note that the same numbers appear in this application of the cosine formula as in Example 13—as indeed they must, since it is applied to the same triangle. But they appear in different positions in the formula, and sometimes with different signs. This demonstrates how easy it is to make an error when applying the cosine formula, and how important it is to put the right numbers in the right places. The formulation in words—'the sum of the squares of the lengths of the two adjacent sides *minus* the square of the length of the opposite side, all divided by twice the product of the lengths of the two adjacent sides'—is very useful in this regard.)

The angle at D in triangle ADB is

$$180° - (\text{angle } A + \text{ angle } B)$$
$$\simeq 180° - (56° + 58°) = 66°.$$

The sizes of the angles of triangle ADB are all close to each other in value (around 60°). This is to be expected, because the lengths of the sides are all fairly close to each other, which means that the triangle is close to being equilateral. Note that the largest angle (at D) is opposite the longest side of the triangle, as you would expect. This provides a useful check of the correctness of the calculations.

(b) A scale plan of the garden, with the angles needed to draw it marked on it, is shown in Figure 74.

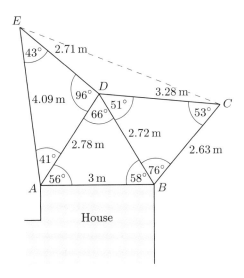

Figure 74

Activity 29

From the information in the right-angled triangles ACD and BCD,

$$\tan \alpha = \frac{h}{x + 1000} \text{ and } \tan \beta = \frac{h}{x}.$$

When α and β are known, these equations each contain two unknowns, x and h. Using algebra, x can be eliminated and hence the value of h found.

Activity 30

The Hollins Cross grid reference is 136845.

To calculate distances from grid references, use Pythagoras' theorem on the differences between the eastings and northings and then scale appropriately (see Chapter 6.2 of the *Calculator Book*).

Mam Tor has grid reference 127836.

Difference between eastings $= 136 - 127 = 9$.

Difference between northings $= 845 - 836 = 9$.

Distance is $0.1\sqrt{9^2 + 9^2}$ km $= 1.3$ km (to 2 s.f.).

Activity 31

This representation of the balloon journey stresses position on the ground and ignores height (and the fuel required to gain height). It stresses certain places over which Chris flew and ignores others. It also ignores the magnificent views, the feelings of the people, and so on.

Activity 32

Leg	Grid ref. 1	Grid ref. 2	Distance east	Distance north	Distance as crow flies/km	Time taken/h	Speed/km/h (to 2 s.f.)
1	123825	157885	$157 - 123$ $= 034$	$885 - 825$ $= 060$	$0.1\sqrt{1156 + 3600}$ $= 6.896$	$46/60$ $= 0.767$	9.0
2	157885	184927	$184 - 157$ $= 027$	$927 - 885$ $= 042$	$0.1\sqrt{729 + 1764}$ $= 4.993$	$27/60$ $= 0.45$	11
3	184927	193940	$193 - 184$ $= 009$	$940 - 927$ $= 013$	$0.1\sqrt{81 + 169}$ $= 1.581$	$9/60$ $= 0.15$	11

Activity 33

Grid ref. 1	Grid ref. 2	Distance east	Distance north	$\tan\theta$	Grid bearing (to nearest degree)	Magnetic bearing (to nearest degree)
123825	157885	$157 - 123$ $= 034$	$885 - 825$ $= 060$	$34/60$	$30°$	$35°$
157885	184927	$184 - 157$ $= 027$	$927 - 885$ $= 042$	$27/42$	$33°$	$38°$
184927	193940	$193 - 184$ $= 9$	$940 - 927$ $= 13$	$9/13$	$35°$	$40°$

Activity 34

(a) The angle between the longitudes

$$= 119.9° - 98.1° = 21.8° = \frac{21.8\pi}{180} \text{ radian.}$$

(b) Substituting $R = 6371$ km gives

$$r = 6371 \cos 25.5° = 5750.370847 \text{ km.}$$

(c) From $l = r\theta$,

$$l = 2187.911318 \text{ km} \simeq 2188 \text{ km (to the nearest km.)}$$

(d) The great circle distance (using the calculator program) is 2185 km.

This is about 3 km less than the distance flown along the line of latitude.

Activity 35

Being able to monitor and evaluate your own learning is an important aspect of being an independent learner. Assessing your own skills, strengths and areas for improvement at regular intervals is a way of monitoring yourself; using feedback from others is another way.